Regaining body wisdom
A multidimensional approach

Silvia Casabianca, Reiki Master

Eyes Wide Open, 2008

First edition, 2008
Copyright © 2008 by Silvia Casabianca
Cover photograph by Sandra Silva
ISBN 978-0-6151-9403-5
www.silviacasabianca.com

All available in Spanish as
El Fin de la Enfermedad. Otra visión del cuerpo humano
Published by Fundación Arte y Ciencia (2005)
Translation by Sandra Silva and Silvia Casabianca
Editing by Carlene Thiessen

Excerpts from this book have already been published by the
Naples Sun Times and the Marco Eagle.

Disclaimer: The information provided in this book is mostly
based on personal opinions and experiences of Silvia
Casabianca, unless otherwise noted.

Advise offered is meant to help readers take informed decisions
and not to replace medical care by a qualified practitioner.

*I wish to dedicate this book to all those
who have served as messengers of the Universe
and turned my journey into a never-ending
learning experience. And also to the
many people who have touched my life.
It would be impossible to make a fair list
without inevitably neglecting someone.*

*Very special gratitude to Sandra, my daughter,
for the time she dedicated to this book.
And heartfelt thanks to Carlene Thiessen
who traded her time editing
this book for my joy translating
and listening to her songs.*

CONTENT

The Awakening 1

  Three stories 1

  Bitacora...7

The New Perspective 11

  Healing does not occur in the blink of an eye 13

  Dear inner healer 17

  Nutrition and the *inner healer* 19

  To heal is to find the lost integrity 22

  Curing vs. Healing 25

  The paradox of progress 27

  Molecules as messengers....................................... 32

  On the health of our planet 38

  Noxious agents – sickening stress 49

  Chemical stress: pollution of water, air and food 52

  Electromagnetic stress: Low frequencies 55

  Emotional stressors, very real................................. 56

  Exhausting thoughts 57

  Testing balance 59

  Life as a luminous halo.. 62

  Communication is vital for life processes 67

  Everything vibrates... 71

  An energy network within a multidimensional body 79

  Binomial of structure and function 88

The Body Wisdom 95

  I. Tegumentary system: our shield 96

  II. Connective system: wrapping and binding 97

  III. Musculoskeletal systems: structure, movement
    and connection 100

  IV. Circulatory system: transport and perfusion 102

  V. Respiratory system: O2 for internal combustion 103

  VI. Digestive system: recycling, storage, transformation
    and disposal 106

  VII. Excretory system: waste disposal and detox.........109

VIII. Reproductive system: perpetuating life 110
IX. Nervous system: evaluation, relation, response,
    regulation and connection.....................................112
X. Endocrine system: regulation, balance, connection 119
XI. Immune system: evaluation, reeducation, defense,
    regulation and connection....................................120
And therefore, organs communicate 123
The *inner healer* responds to the challenges imposed
by the environment..127
Inflammation and reparation 128
Pain and pain killers 132
Health and illness 136
Inquisitive minds formulate new theories 144
The body speaks to us..147

Reiki and the art of healing........................ 151
Founder, Mikao Usui (1865–1926) 153
Reiki precepts 156
Initiation to Reiki..158
Reiki seminars 161
Reiki practice..165
The master 167
On becoming a healer..169
Know thyself..171
The healer as apprentice 175
The therapeutic alliance..177
The healer seeks a rhythmic life 179
Primum, nil nocere 181

Annexes.. 185
Annex 1. Research and Reiki 187
Annex 2. Nutrition, a fundamental step to health.......190

Bibliography... 197

All truth passes through three stages.
First, it is ridiculed.
Second, it is violently opposed.
Third, it is accepted as being
self-evident...

Arthur Schopenhauer

The Awakening

Three stories

1. February, 1993

Master Tito[1] spoke with ease, offering explanations that would have elicited reservations in any orthodox scientific mind. A summer ray of sun shyly peeped through a door that was ajar and his voice echoed in the gloom. By visiting him – an unusual kind of medic – I was crossing a threshold and questioning the *infallible* science I learned in medical school that had guided my practice for many years. He made me wonder.

What is science, after all? Any dictionary would say science is knowledge attained through study or practice; and I will add that knowledge is not necessarily attained in academic settings. Maybe in our quest for truth we could take other paths that do not necessarily go through experimental verification.

The consultation with Master Tito was actually for two friends who were searching for a solution to a fertility problem. They had been told that a tumor in her thyroid gland might have been the cause of two former miscarriages.

---

[1] Names have been modified to respect the privacy of all persons involved in the stories.

The healer's conclusion did not make sense to me: "Infertility due to 'cold womb[2]'"? Even though I tried to stay open-minded to new ways of explaining the human body, this remark put me on the alert. Was he a *charlatan*?

From time to time, the manly figure with dark eyes and beard blurred in front of me and then I could perceive a subtle light surrounding him. Every time I observe this glow around other people I think it is merely an optical effect prompted by a meditative state. Actually, I doubt it isn't. In his case this luminosity seemed an atypical *aura*,– it extended a few inches off and around his body and included a wing-like extension on the left side, about one meter long.

A self-proclaimed clairvoyant, Tito was emphatic about his diagnosis. My friend did not have any disease related to her thyroid gland, and the infertility problem would go away by taking some herbs, practicing some exercises and making nutritional changes in the couple's diet.

My friend and her husband followed his instructions closely and a year later, their first offspring was happily kicking around.

2. April, 1993

Master Elias stood in front of the room talking about things that to both my daughter and me sounded like either a joke or a folly. Even though he had not yet reached age 40, his beautiful long beard had turned gray, giving him a patriarchal image. His long and curly hair hung with the same grey undertones. His inviting smile and intense words captured my attention in spite of the apparent nonsense of some of his talk.

Master Elias belonged to the same group as Master Tito but he had actually graduated from a renowned school of medicine in Colombia. He was dedicated now, and for some years already, to another path he called *Universal Medicine*. His speech had such an intrinsic coherence that it could only be questioned within its own paradigm.

"I am a messenger angel," he announced at the beginning of his speech. He then explained that there are three types of medicine: abysmal *medicine* that invades, intoxicates, amputates and destroys; earthy *medisine* that lost track when alcohol started to be added to preserve the energetic principles in which it is based; and *Zelestial*

---

[2] Only recently I learned that the term "cold womb" is actually used in Traditional Chinese Medicine. I have no idea where Master Tito learned this concept but I am sure he had no formal training in any health care modality.

medizine that he said he had come to reveal to all of us, which respects the body and its self-healing power.

Some of what he said clicked in my head, enough to stay over for dinner; however, I did not find a way to formulate the questions I had in my mind. We ate in a vegetarian diner next door to the 'wisdom school' where people from all paths of life came to listen to the talks. A few of them were the Master's patients.

The school wanted to found an academy of alternative medicine and the Master was looking for basic science teachers. That's why I was there, captivated by the invitation to be a part of such an innovative project and attracted by the people's kind acts and respectful manners.

A young lady, who was a member of the youth group I mentored back then, also attended the lecture and became my reason to come back to the place. She wanted to invite the other group members to listen to the Master and I wanted to know what he had to say to *my kids*. With all the news on deceitful cults, I felt it was my responsibility to check this out first hand.

Master Elias organized a trip to the botanical garden in which some of the youth took part. On the green lawn, under the hugging shade of enormous ceiba trees, the Master offered his version of the creation of the world. It was a mix of Confucian, Buddhist and Taoist Chinese mythologies. The youngsters listened and questioned him with a hint of cynicism and much skepticism. The Master however had answers for everything and no question took him aback.

On my way back to the car, I asked the Master for a "prescription," made of herbs or whatever he was using that seemed to have so much success with his patients. Just like that, I thought, from one colleague to another, no need for more protocols. My health was not the best, I told him and he breathed a sigh of relief. He had noticed my *low energy*, he replied, but he had to wait and leave it up to me to ask for help. The Master requested that I come in for a consultation.

The concept of healing by the *laying on of hands* was new and foreign to me. I did not know anyone who had experienced such a thing. It seemed to me that those persons who consulted the Master and talked about their significant improvement were mostly enjoying the benefits of his non-pharmacologic prescriptions. My orthodox medical mind was still predominant. Did I want to consult him because of a curiosity attack or because my soul needed to know and be able to believe in that which is intangible?

In their school's backyard, under the shade of a mango tree, the master made me stand with hands in prayer position. I followed his instruction to close my eyes and then felt his hands around me without touching. I thought I heard him whisper something. Minutes later, I opened my eyes and he explained to me he had just done a *laying on of hands*. I slightly shrugged thinking "What's the big fuss?" I was still looking forward to receiving my prescription. He then proceeded to do a *pendulum reading* using a pink quartz crystal and his clairvoyant gift.

His evaluation of my body was alarming. Energy was too low in all my organs except kidneys. "Let's now take a look at your nervous system," he said while pulling the thread to stop the pendulum from moving. He focused. A few seconds later, the crystal started drawing small circles around the palm of my hand and then stopped suddenly. Just like that, organ after organ. I had not yet confessed my cigarette addiction but my lungs did not pass the exam; the pendulum hardly moved. At the end, after noticing I had millions of queries, he showed me the movement of the pendulum on his own thigh. The crystal moved rapidly and drew circles of about 15 centimeters in diameter.

I looked at my watch and started heading out. I had a meeting at the university where I taught. I said I'd come back in the evening for the prescription. He took some notes and we said good-bye.

My diet was rather poor in those days. I was exhausted all the time, I had frequent headaches and my digestion was a disaster. Some canned peas – to pretend I was not missing the required vegetable servings – whole wheat bread and, once in a while, a banana or cereal, made up my diet. All these complemented by lots of black coffee and smoke - cigarettes before and after each meal, before going to sleep and as soon as I woke up, after the shower, while I was talking on the phone The only time I did not smoke was when teaching or in a closed-door meeting. And I still had the nerve to ask myself why my body was not functioning properly.

After lunch that day, I routinely opened my purse and when I spotted my cigarettes, I realized with surprise that I had not smoked even once during that morning. Maybe it was just that I had been very busy. Around 5 o'clock that afternoon, I received a phone call from my colleague, the one who introduced me to the master. "How was it?" she asked about the consultation. It was at that moment that I realized I had not smoked since that morning and did not feel like smoking even now.

When I returned in the evening of the same day to pick up my prescription, the master asked me, "How are you feeling? What have your thoughts been like?"

My answers where condescending. How was I supposed to feel? Why would I have to have any particular thoughts? I had not even received the actual prescription yet. Would any change take place without the help of a chemical product? My medical mentality was overpowering my thought process once more. However, his third question left me astonished. "And, what about the cigarettes?" OK, it was possible that my friend had told him I had not been smoking that day, but I later confirmed she had not talked to him.

The Master then gave me some metaphysical explanations that I barely wanted to hear – that he was an angel sent to perform this miracle, that I needed to be healthy in order to achieve my mission, and he was referring to a brand new project of alternative education that was an extension of our successful experience with youths.

I requested my prescription somewhat impatiently. I listened to his recommendations. I accepted his suggestion to try an alkalizing vegetarian diet for three months.

It has been 16 years since and I have never, even once, felt the craving for a cigarette again. Moreover, I was glad to see how the diet radically transformed my body and my health.

How does the *smoker's memory* of someone dominated by an addiction of 25 years get suddenly and forever deleted? How does an addiction to cigarettes stop without the anguish and anxiety of withdrawal, one that I had already experienced in the past when trying to quit? My inquisitive mind was rebelling. I searched for answers in the master himself but found myself trapped in his self-contained belief system.

Could I learn to do for others what he had done for me? Of course, he told me. The requirement? Only to be perfect! Ha! This answer killed my interest for the moment. I was not prepared or ready for the challenge he was proposing. Who achieves perfection?

I learned the exercises that his group practiced and my body gained strength. I followed their diet and my body gained more strength. However, I could not find with them the spiritual life I was searching for. As a lifelong apprentice I was ever interested in exploring new horizons.

A couple of weeks after the consultation, the master sent me a small bottle of Bach flower essences that helped me deal with any residual anxiety that manifested after meals. I wondered what these essences were and I found the principles of homeopathy. An energetic principle of a flower contained in water? My old paradigms kept cracking before me.

In the meantime, everyone started talking about my de-aging process. I exercised daily. I felt rejuvenated and energized and I began finding serenity within myself in places that I had never intuited before. I was so puzzled that I busied myself visiting libraries and book-stores and I submerged myself in another world of unexplored depths: the universe of energy medicine.

3. June, 1994

I've heard on numerous occasions that the master comes to you when you are ready. I believe we are all masters of each other and that, in the long run, the goal is to find our own inner master. My daughter has been my great master in an infinite number of matters, for example. In the specific case of reiki, my masters have appeared not only when I needed them, but also when circumstances favored my learning from them. That is just how my master Maria came to me a year after I began the healing process, and I decided to initiate myself to reiki with her.

I met Maria at a vegetarian restaurant in downtown Cartagena where it always smelled like good homemade cooking. I was introduced to this slender woman of expressive eyes and long, black, curly hair. Maria Adelina Sastre, from Spain, reiki master. I had no idea what reiki was; I had never even heard of it. Maria had lived in the city for a short period of time and she was putting together an informative gathering for the following day.

That Friday night I was there among strangers. As usual, I was the most curious and I once more forgot about the intimidating power of my questions. Maria was a teacher with obvious charisma, humility and serenity. She explained to us that reiki was made up of two Japanese words: Rei, meaning universal energy and Ki, referring to the energy that flows within our bodies. When Rei and Ki flow in harmony our physical, mental and spiritual health are optimal.

Before making up my mind about taking the reiki class with Maria, I met with a distant relative of mine who I found out was practicing reiki as well. I wanted information and she knew that one session was more powerful than words. I had overblown expectations based on what I knew about *universal medizine, which* the master practiced. I did learn that with reiki there are no diagnoses and you do not direct the energy.

After a few minutes into the reiki session, my body felt like it was sliding though a tunnel as I entered into a deep relaxation state. That was all. Nothing out of the ordinary, I thought. My curiosity was intact but once more I had felt the effects of something I could not explain. And I wanted to know more.

I took reiki level one with Maria. I let her initiate me to reiki during a weekend, in a small room in the back of a busy hair salon, with smells of incense and oils, surrounded by relaxing music. I was one of only two students taking the class.

During the brief moments that the first of four initiations took, while I was sitting by a window ready to receive the symbols that would be inscribed on my aura, my body was filled with light and became so weightless that it seemed it would levitate at any moment. I suddenly felt an immense happiness and peace.

It was a new beginning. Neither my life nor my concept of the human body, of health and illness, would ever be the same.

Bitacora

Before I sat to write this book in Spanish, I shared my new ideas on health and illness with a former colleague and life-long friend. This is the almost literal content of the letter I sent to him.

"There is a phenomenon nowadays that I believe will transform the medical practice forever. Unfortunately, it won't be to stop privatization of medical services but at least to turn us into individuals responsible for our own health. I am referring to the rupture of the dependency from the 'expert,' which suggests to me a very interesting tomorrow. The increase in medical costs and prescription prices, the medical errors that take so many lives, the awareness of the importance of our active participation in the medical process, have been bringing us to the conclusion that there is no better expert nor anyone who knows more about our own body than ourselves. We just need to tune in.

Many names: biological medicine, alternative/complementary medicine, integrative medicine, holistic medicine, era-three medicine, vibrational medicine. They all come up when speaking of shifting perspectives on the concepts that have prevailed about health and illness in the Western world in the last three or four centuries. Ever since we started having access to the East and ancestral practices such as acupuncture became available for general scrutiny, various paradigms have competed for the acknowledgement of being pioneers in the interpretation of the subsequent discoveries or the only ones privileged owners of the truth. However, when exploring those paths, it is obvious how much they all have in common, in spite of the different wording their theories use.

Different publications report theories that confront truths we have taken for granted for centuries. The 2003 issue of Scientific American *included the article* Is tri-dimensionality real or an illusion? *regarding the holistic theory applied to the Universe. Rupert Sheldrake explained during the 1990s the existence of a morphogenetic field on which our physical structure would be designed. Author Fritjov Capra talked about the immune system as an inner communication net that he also calls our second brain. The* Chaos Theory *shows us the interconnectedness within the universe manifested in a series of infinite subtle relations, and calls us to be co-responsible individuals for everything happening now and tomorrow.*

The Scientist *published in 2004 a brief study,* 'Longevity gene, diet linked' *by David Secko, that confirms that the type of diet interferes with or stimulates the manifestation of certain genes in rats. It actually revives old Lamarck's theories that had been displaced by Darwinism! Electrical appliances, magnets and lights are commercialized to be used as anti-inflammatory devices and non-invasive alternative treatments are proliferating. Research has confirmed that reiki accelerates healing of a wound, stimulates the production of white blood cells and speeds up coagulation times.*

New paths for science have been born even if at a slow pace. However, unlike 'official' science, there is not enough governmental or private funding for research and implementation of alternative therapies. Up until now, only products with a commercial profile have been supported. That is why, while some compete to mutate genes in the labs, to clone organisms or treat diseases with stem cells, the new medicine remains close to the simplicity of former truths such as Hippocrates' food-shall-be-our-medicine, we shall breathe in pure air, we shall stay active and restore the lost balance, and lastly, stimulate our inner healer.

Today's science does not seem interested in following the reasons why any given Joe starts a spontaneous remission of a terminal cancer after going through an introspective process about his limitations to love and forgive. Or the reasons why a person with AIDS has his symptoms disappear after he suspends the use of antiretroviral treatments and changes his lifestyle in a radical way. Today's science, with due exceptions that confirm the rule, seems to focus on an arrogant race to provide health care professionals (as if they were some sort of mechanics) with the means – that is, the power – to control the patient's body and 'repair' what's been broken.

Perhaps the tendency toward privatization of the current health care system – so familiar to us – may be in fact only a symptom of its agony. Arthur Hailey's Strong Medicine *published in the 1970s – well-documented fiction – alerted the public to the risks of scientific research led by pharmacological laboratories. He offered a picture in which medicine borders scam, and service quality depends on the need to reduce costs in order to guarantee inflated profits. This type of medicine cannot do other than perish, in spite of the investment big corporations have in it. Alternative practices succumbing to the same commercial vice will not make it either, in just a matter of years.*

As long as we understand that our health depends on our lifestyle and we become more responsible for our own well-being, medicine and clinical institutions will have to search for solutions other than prescriptions and surgery.

I feel we should try to see beyond our inherited dogmas, those that we tend to live as unquestionable truths, even if that requires us to confront them and leave them behind."

The new perspective

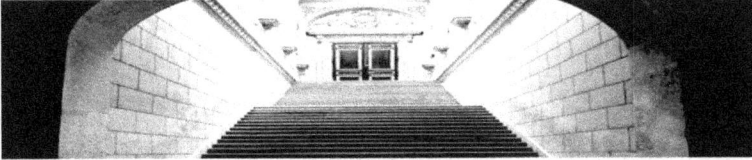

"Knowledge, like wealth, is intended for use"
Kybalion

I have had the privilege of initiating around three hundred people to reiki. Checking with other reiki masters, I have found that usually, only a small percentage of students go on to seek the second level and just a few work toward the third level or master level. I wonder why, if the number of people who are initiated to reiki multiplies daily, there are so few of them who incorporate it into their daily lives or professional practice. Some use reiki only with themselves or with friends, relatives or pets. The ones that practice reiki professionally – doctors, nurses and massage therapists among them – introduce it rather shyly to their patients or add it as a complement to another treatment. Notwithstanding, almost everyone agrees that reiki has been a transformational and positive experience in their lives.

Something else that has caught my attention is that some of these reiki practitioners who have reached the second and third level, limit themselves to apply what they learned in reiki level one. They find it difficult to memorize and to understand the symbols learned in the other levels and they feel a hint of hesitation when using them, as well as to letting intuition guide them.

There may be many explanations for this. I've heard masters say that classes should not be too short or that they should be a bit more demanding, because apprentices do not receive enough information. Other practitioners deem that the public in general does not have enough knowledge about reiki and doubt it can be accepted. However, I have seen how reiki has become a subject of an increasing number of television and radio shows, and press articles that talk about it are numerous. I also know that more people have become receptive to energy healing and have decided to experience reiki thanks to this media promotion. I make out that some students have been disappointed because they expect faster and more dramatic results in cases of physical illnesses. Some others get anxious to respond to the expectations of those to whom they offer reiki.

It seems to me that another plausible explanation for this limitation in the practice of reiki is the prevalence of certain beliefs about health and illness, healing and curing, which are based on allopathic (conventional) medicine, concepts that we have not yet questioned enough. These beliefs don't allow reiki practitioners to use reiki in all its simplicity, effectiveness and beauty. They also make those who come for reiki expect what they are offered in conventional medicine – a rather quick fix for what according to them is broken and an expert who will perform the fix, since they don't recognize their body's self-healing capacities.

I frequently hear reiki practitioners say they have not studied enough, practiced consistently enough or that they still need to experience the so-sought-after transformation and thus they don't feel sufficient merit to heal others. However, when they reflect on it, they discover that there is a key difference between mere accumulation of information acquired through classes and readings, and the knowledge that stems from contact with another, being present and developing deep intuitive perception. In order to develop their capacity to sense and feel, it is not information but rather the knowledge obtained through practicing reiki that they need. In brief, they are denying to themselves exactly what they need: practice. For it is practice itself that will teach them what reiki is and how to use it.

The *I-am-not-worthy* attitude that may tend to reflect humbleness is in fact a subconscious expression of insecurity and arrogance (opposite sides of the same coin) because we are not doing the healing; it is the universal energy of which we are just channels, it is the body responding to this energy. When students who receive a reiki initiation understand that they are not the ones performing the healing, they conquer resistance and incorporate reiki into their daily lives and professional practice as a tool that enlightens, harmonizes, raises consciousness and calms them down. As reiki practitioners there is not a special power that we own, but a gift we have been privileged with which allows us to tap into this source of immense universal, divine, energy and apply it to a body that is already equipped with the intelligence to heal itself. This energy and the way we become capable of funneling it is totally beyond our rational understanding.

12

Everyone can learn reiki, everyone can benefit from it. There are no pre-requisites to practice or to receive reiki. That is why I insist that all of us reiki practitioners have the duty and commitment to make it available as a natural healing method to all who seek and accept it.

One of the axioms of the *Kybalion*[3] says,

> *The possession of Knowledge, unless accompanied by a manifestation and expression in Action, is like the hoarding of precious metals – a vain and foolish thing. Knowledge, like wealth, is intended for Use. The Law of Use is Universal, and he who violates it suffers by reason of his conflict with natural forces.*

Why would we want to bury this precious metal? Let's incorporate reiki into our daily lives. Let's get rid of obstacles that keep us from practicing it; let's revise the concepts that keep us stuck. To delve into the knowledge that keeps us from practicing reiki, I propose another type of understanding that proves the body's capacity to harmonize with the universe and heal itself.

Healing does not happen in the blink of an eye

For years, we have been used to looking for the specialist, wanting to experience an *aspirin effect*, fast and affordable treatments that involve no effort, or looking for an immediate cure. It is difficult for us to believe in treatments in which the practitioner is not an *expert* but a simple instrument (of the universal energy) who is guided by intuition. It is also difficult to accept that true healing is not a miraculous vanishing of a symptom, but instead a process that requires our active participation, our change in perspective, our will power and the rising of our conscience as the result of our reflecting on the why and the purpose of our symptoms.

We are so used to revering medical doctors that it is difficult for us to identify with the image of a humble healer who, using her intuition, pauses to feel, offers complete focus, trusts the body's self-healing power and knows that the disease is an expression of an imbalance. She also knows that and the symptom is a mechanism of the body to try to restore balance, and a calling of the body for us to introduce changes that require time.

---

[3] Kybalion: said to be the essence of the teachings of the Greek God Hermes Trismegisto. The book was published anonymously in 1908 by a group or persons under the pseudonym of "The Three Initiates".

In a world where people are valued mostly for their academic titles, intuition does not have the authority that college credentials offer. We miss the fact that healers usually invest as much time as scholars do, not necessarily browsing libraries but in the quest for awareness, constantly observing their inner lives and practicing the ability to be present. And we dismiss the therapeutic worth of something as simple as an authentic presence that facilitates the necessary intuition to support the other person's growth process.

The prevalent way of thinking in a society is omnipresent; it scatters and reproduces itself through education, advertisement, culture and stereotypes, validating and stating as obvious some *truths* that must be accepted as if there were only one right way to do things. We subconsciously absorb these ideas and fail to assess and question their validity. Those ideas have impregnated us and we end up defending them as if they were our own sacred truth.

We have been directed to fragment the body, divorcing it from the mind, and to label symptoms in a packet we call diagnosis, so we can prescribe a treatment. We erect an identity around affirmations such us: "You have allergies," "My ulcer," "I have arthritis." And others reinforce this "sick person" identity by saying things like, "She cannot do it because she is/has…"

When studying illness, medicine tends to stick with the apparent causes and the prescribing of symptomatic treatments. It may seem obvious to treat an infection with an antibiotic or even a homeopathic remedy. The problem is that with such treatments, in the long run the patient identifies herself with the symptom. We hear people saying, "I'm suffering asthma, meningitis, or migraines…" This lengthens the duration of disease and generates a codependent relationship in which the doctor saves the helpless patient. On the other hand, we rarely hear things such as, "Every time I eat too much candy, my immune system weakens and I become more susceptible to germs or allergens."

The ideal for us therapists, in order to establish a true healing experience, would be to help the patient go through an integrative process, meaning the raising of our consciousness and the accepting of that part of us that we have been denying.

Back when I practiced as a medical doctor, I used to see a little girl every month. Her parents would bring her with a different infection each time: throat, ear, bronchi. Twenty years ago, the concept of a depressed immune system or of the relationship nutrition/immunity was not yet popular. Since the girl's infections were almost always due to streptococcus, she received an injection of penicillin time after time in order to prevent rheumatic fever – a potential complication of these infections – and she also had frequent blood tests to make sure the risks were under control. My insistence on introducing nutritional changes did not convince the mother, who thought it would be more difficult and upsetting to make her daughter quit the *junk food* that comprised almost her entire diet. Without a change in the pleasing-mother's role and without a change in the diet, true healing could not happen.

When we finally acquire a different comprehension of what health and illness are, alternative healing techniques such as reiki will expand and benefit all. A paradigm shift to accepting the body's capacity to regain its balance is needed; it must add force to a non-invasive approach that respects the human body. We will then understand that there is no such thing as *illness*; that symptoms are an expression of an imbalance making the body's alarm system turn on. Symptoms are the body's attempt to get rid of whatever is harming it, to protect tissues from ulterior injury and to adapt to new conditions.

It does not matter what the condition is or the expectations of the one receiving the treatment. Reiki will always be administered in the same way, aiming to unblock the energy flow and stimulate the body's own intelligence and capacity to heal itself. The effects could manifest immediately or in the long run, and in every dimension[4].

Reiki treatments are usually not limited to *laying-on of the hands*. Practitioners suggest that receivers adopt a simple lifestyle, less automatic, in which they will opt for that which is healthy before that which is fast, and invite recapitulating about a few guiding principles. Reiki does little for reverting a physical condition if there is not a will to change what set it in place, if the body has almost completely lost its ability to keep a certain balance, or if the person keeps on being exposed to toxic and stressful factors.

Paraphrasing Francis of Assisi, reiki offers us the serenity to accept the things we cannot change, courage to change the things we can, and wisdom to know the difference. It has been my experience, and that of many students and practitioners, that reiki is beneficial in every situation, even if it is not clear to us what the outcome might be.

A few years ago, I had a case that illustrated the mysterious ways in which reiki contributes to healing. A 44-year-old patient was referred to me due to abundant uterine bleeding caused by fibroids. She

---

[4] See table 2, page 81.

refused to have the surgery recommended by her doctor. She had heard about reiki and wanted to try it out. During our sessions, Marina[5] explained to me that she had never had children, but even at her age, though she was not involved in any relationship, she had not lost the hope of becoming a mother.

Marina revealed that toward the end of her adolescence, her boyfriend had abandoned her on the eve of their wedding, which had caused her a great sorrow that ended up isolating her from the world for a long time. When, years later, her father – whom she adored – died, her earlier non-resolved loss got in the way of processing her new sorrow. In the days I was treating her, the relationship with her mother was not harmonious, as she wished it were. After the third session, Marina felt the need to get close to her mother and expressed her love by hugging her, which she had not done in years. This surprised her mother who responded positively. Therapy extended several weeks beyond the four initial sessions that constitute a complete treatment. Her bleeding stopped completely and she went back to her hometown. Two months later, when she felt stronger, Marina decided to have the hysterectomy. Processing her previous losses in the reiki sessions allowed her to face the loss of an organ and her hopes of maternity.

Because reiki is practiced with a holistic perspective that seeks healing, the approach remains the same regardless of the person's problem, symptom or diagnosis. However, uniqueness is valued because the therapist *feels* the energy in different parts of the body and uses her hands accordingly in order to transmit the universal energy that she has funneled. Moreover, each word, each gesture, each glance, offers information that is used to increase the receptor's consciousness, a pre-requisite to stimulating the changes the body needs.

The word therapist comes from the Greek language and means *assistant who serves the gods by following a ritual*. Before Hippocrates introduced *home visits* to the practice of medicine, patients went to temples seeking the gods' help in finding the meaning of their ailments. Therapists were mediators in this process. A reiki practitioner is a mediator too, a facilitator to who people come in search of help, not miracles.

During every session reiki practitioners administer, they get in tune with the Universe's energy, which helps them increase their

---

[5] Name and details have been modified to protect confidentiality.

consciousness and develop their intuition. There is no way to become a healer other than laying on of hands on yourself and others.

I trust that reiki practitioners will be more pro-active in spreading this healing method as they fathom in a multidimensional perspective of the human body and then assume their commitment to utilizing the knowledge they have acquired. This way, they will honor Mikao Usui's goals when he created this method. He said, "I would like to make this method accessible to the public for the well-being of humanity. Each of us has the potential to receive a divine gift that results in the reunification of the body and the spirit. This way [with reiki] a great number of people will experience the blessing of the divine."[6]

Dear inner healer

Mainstream Western medicine is starting to pay heed to a principle that has been endorsed by other cultures: *Our body is a self-regulating organism that contains an inner healer.* This healer, which is not limited to instinctive reactions as it has been understood under a Newtonian paradigm for the last few centuries, is in charge of surveillance and communication, storage of information, evaluation of what is going on in the body at any given moment and organization and expression of the body as a whole.

The *inner healer* is also responsible for providing suitable solutions to adaptive challenges imposed by the environment. It draws on information the body has memorized and learned in order to perform its functions and therefore, we can call it an intelligent healer.

Each one of our skin cells lives for about 36 days. When one cell dies, another replaces it. How else could we explain that our skin lasts a whole lifetime?

Our red blood cells live up to 119 days. However, the number of red cells remains constant in the blood.

We completely renew our body tissues every seven years. It happens without our intervention, although, of course, we need to guarantee the *raw material*. Because our tissues are made up of materials that come from the nutrients we eat, the water we drink and the air we breathe, the quality of our tissues will depend on the quality of our food, water and air.

---

[6] From the book *The Legacy of Dr. Usui* by Frank Arjava.

Who instructs our body to do the regeneration and repairing jobs? How does the body know that it has to build skin cells in the skin and red blood cells in the blood?

We have to assume that there is intelligence imprinted in our organism. There is some sort of software in our energetic (subtle) bodies and in our genes as well, that maintains our life. Some kind of blueprint within our cells must mediate the communication system in the body, granting regeneration and reparation of our tissues, and therefore, survival.

Stress is what defeats this *inner healer*, which resides in our subtle bodies as well as in the depths of our entrails; stress breaks the balance and generates *dis-ease*.

Stress is the result of the lifestyle we have chosen according to cultural, social, and financial factors. Among these are the roles that we play in society, the quality of our interpersonal relationships, the preference for processed food over natural produce, our nutritional habits, the way we exercise and breathe, our self esteem, our sense of safety, our spiritual life and our positive or traumatic experiences. All these elements affect the way in which our body responds to stress, which is a constant in our ever-changing lives.

A certain amount of stress in life is unavoidable and even stimulating and healthy, and the body is fully equipped to deal with it. However, excessive stress has a cumulative effect that ends up compromising our body balance, hindering the body's capacity to respond to stressors. Our capacity to respond to stress varies in each state of our life cycle, weakening us or helping us develop resiliency[7].

Health professionals who have chosen to practice in the fields of family medicine, public health and rural medicine know well the role that lifestyle plays in maintaining health. This is also well known to refugees and displaced people, populations affected by violence or disasters whose most common ailments won't probably show up on x-rays or MRIs because they are just the result of mounting stress, deprivation and detrimental life conditions.

---

[7] This term was initially applied to the characteristics of certain metals that could bounce back to their original form after being deformed. Ecologists also use the term to refer to ecosystems capable of maintaining diversity and integrity after suffering certain disturbances. In general, it is used to define the learning achieved and the capacity of an organism to recover after a traumatic event, allowing it a level of functioning even more competent in adverse conditions.

At the same time that technology feeds our ability to wonder, old basic truths about health and illness are resurfacing and being endorsed by scientific research. These truths speak of the human body as a marvelous system of systems, multidimensional (physical, emotional, mental, spiritual, social and cosmic) with an immense capacity to preserve, regenerate and repair itself.

How long will we continue to deceive ourselves by accepting a medicine that forces the laws of nature? Why continue in this path if there is clear evidence that by modifying our nutritional habits, exercising, reducing toxicity and stressors, we can in most cases avoid disease or keep symptoms at bay?

Nutrition and the inner healer

In August 2003, the *New Scientist magazine* published the article *You are what your mother ate* by Philip Cohen. It reported a study proving that baby mice would change skin color if their mothers were fed different levels of common nutrients during their pregnancy. The babies with the darker skin tone were also less susceptible to obesity and diabetes.

The relationship between heritage and nutrition opens exciting possibilities. A resulting new field is that of nutritional genetics or nutrigenomics (not yet in our dictionaries), which can examine the byproducts of metabolism and use informatics to identify and predict what impact nutrition will have on the health of individuals with certain genotypes. Genomics refers to the identification of an organism's sequence of genes and its variants, and nutrigenomics is the specific application of this knowledge to food processing and consumption.

The increased availability of organic products in today's supermarkets is zilch in front of the forthcoming healthy food revolution. Picture a future where food-shelving in supermarkets adjusts to people's genetic types. A future where the general public will still be educated to shun sodas and stick to mandatory five-veggies-a-day, but will have a much better account of why everybody does not respond equally to the same diet.

Nature is controlled by three biological laws, argued Jean Baptiste Lamarck in 1809: environment influences organ development; the body changes its structure according to the use and disuse of its parts and these acquired characteristics can be inherited. After Charles

Darwin explained evolution by natural selection in 1859, biologists discarded his predecessor's ideas.

Recently, however, the German scientist Andreas Plagemann revived Lamarck's theory, concluding that a high diabetes risk can be passed on to several generations, and that this is not caused by spontaneous mutation, but rather is due to the inheritance of an acquired condition. Studies have shown that unborn rats of diabetic rat-moms have increased levels of insulin. It seems that certain brain cells (the ones in charge of hunger and satiety) are irreversibly damaged by the excessive sugar in the mother's blood[8].

Plagemann has not been alone. How to elucidate, for example, why lactose intolerance affects Asians and Africans more often than northern Europeans? According to biologist Jim Kaput, founder of the diagnostic company NutraGenomics, this is explained by the fact that, between 6,500 and 12,000 years ago, a change in Europeans' DNA occurred that allowed them to digest lactose during a season when food was scarce and milk became essential for survival. This modification was passed on to their offspring, he concluded.

The *International Human Genome Project*, formally finished in 2006 after 16 years of research, left scientists with a reference map for the 25 thousand or so genes in the human genome (hereditary information encoded in the DNA) and the more than 3 billion common variants lurking inside those genes.

Why are some people more likely to suffer cancer or cataracts than others? Is it not yet completely clear how the genome would explain health and disease. Researchers try to answer these and more looming questions as they look to the impact diet has on our genes. The work done so far to identify our genetic map carries the promise that labs will be able to provide individual genetic profiles. As scientists achieve an enhanced understanding of the relation between nutrition and genetics, we will be able to eat *a la carte*, according to our individual DNA codes, to prevent and mitigate aging and chronic illness.

Ciao to genetic determinism! Our genetic inheritance won't give us any more troubles. The results of a simple blood test will be sent to us with recommendations on what foods will keep us healthy. As you may imagine, the food industry is getting ready to make its next trillion dollars thanks to this new knowledge.

---

[8] Visit: www.ourfood.com/Nutritional_Genomics.html.

Scientists are also becoming familiarized with the intricate way in which genes operate, with "switches" that turn on or off, depending on the organism's interaction with the environment. All together with the study of how food's bioactive compounds work in the body, more possibilities are opening to the development of personalized diets and medications that will work *with* the *inner healer*.

The turn of this century finds science exploring further into the human organism and forcing it to reveal its deepest biological secrets. In the astounding past decade, technology has provided new tools not only to better scrutinize the body but also to intrude and alter it. Technology applied to biology fascinates and scares me. Humans have gone quite far in the search for knowledge, and maybe too far in their ambition to manipulate and control biology. Instead of acquiescently and respectfully accepting and waiting for nature to do its job, man wants to step ahead and become the He-god of evolution. Nobody knows at what cost.

Nowadays many people suffering from cancer say no to radiotherapy, chemotherapy or opioid pain killers, to avoid detriment of their quality of life, especially when survival will not be significantly improved by the treatment. Interestingly, many of them have reported *miraculous* recoveries that some authors (browse Internet, for example, for the Cancer Report or the New Medicine of Dr. Ryke Geerd Hamer) link to the fact that the inner capacity of the body was respected and when the conflict that generated disease was over the patient recovered.

Health spending is really high for those governments that do not promote healthy living conditions and healthy lifestyles. It would be preferable to focus efforts in health education so that communities recover healthy lifestyles. Money could be better spent in sanitary infrastructure that allows the prevention of most air and waterborne diseases. Prevention in third-world countries should be focused on improving infrastructure. Prevention in developed countries should be based on regulations to the food industry and healthy lifestyles.

For about a half century the World Health Organization (WHO) has been raising awareness about the importance of prevention and primary care. But unfortunately in the United States and many third world countries, the trend in health care is not global coverage but privatization. When health becomes merchandise, profit mandates health policies. The direct result of privatization has in many countries meant the end of prevention programs.

I am sure that finding the path to valid solutions is not far. As a byproduct of the illusory fountain of youth, we have a better understanding about the effects of nutrition, physical activity and stress on our bodies. We can educate ourselves now on how to support the *inner healer* to help it restore balance. This way we can reconnect with our body and our environment, reassuming responsibility for the way we eat, move, breath and respond to stress. We will become increasingly aware of how the body expresses its needs and of the immense capacity of the body to heal itself. We will also beat the restrictions that hinder the realization of our potential and that are a continuous source of pain and limitation.

One of the roles of the reiki practitioner is precisely to stimulate and support the *inner healer*, to help patients rethink the path that has led them to illness and to retake a more harmonic and simple lifestyle.

This book proposes a new vision of the human body, moving away from the Newtonian paradigm that visualizes our bodies as machineries that decompose or break. No! Humans may be repaired but never healed by just replacing parts or *pipeline* pieces.

With all due respect that I have for achievements in the fields of diagnostics and surgery and the lives that have been saved or rehabilitated thanks to the advances in medicine, I feel a sense of urgency that we should move toward visions that are integral and integrative.

I deem it fundamental to recognize the multidimensionality of the body and overall, to recognize that the body is a sanctuary for an inner intelligence that promotes and sustains regeneration and healing. We also need to understand that we are not only biological beings but also social and cosmic beings.

To heal is to find the lost integrity

I was answering questions on reiki during a break at an event when a middle aged woman approached me to ask if reiki could help a diabetic.

"I am a diabetic," she said.

"What would it be like if instead of saying you *are* a diabetic, we forget the diagnosis and say that your body is having trouble handling sugar?" I inquired, as I watched her mentally seeking an answer.

I had in my mind something I'd read about a tribe in Malaysia whose language does not include the verb to be. Thus, they have no way to label people. There is no "You are pretty" or "you are bad" or "you are diabetic." If you were to say, "You are selfish," you'd have to put it this way: "I see you taking care of your needs but not the needs of others."

You'd tell people what you see or how you feel, instead of what they are, which usually instills guilt or helplessness.

In the past, I played with the idea of suppressing diagnoses. With interns at the medical school where I taught, we observed a small group of recently diagnosed patients with high levels of blood sugar but who didn't require insulin.

We empowered the patients by describing their conditions in terms of body imbalances and explaining the ways they could support the body's healing process.

These were more compliant with their physician's recommendations than patients who had received a diagnosis of diabetes, and all went back to normal blood sugar levels in less than six months.

I said to the lady, "When we use the verb 'to be,' we attach an identity to it. When you hear 'You are diabetic,' or you state 'I am diabetic,' you're inserting a label that the body understands as a command. You may continue to be diabetic from that point on. But, let's look at it from a different standpoint, and rephrase the diagnosis. Tell me in your own words what is happening to your body."

She explained that she was hungry, tired and thirsty most of the time and her blood sugar rose to around 200 (normal would be around 120 mgrs/dl).

"Not so high," I said.

"But they make me take medication, and I still feel tired and hungry."

"Depending on the case, some people need medication, but other people have achieved normal blood sugar levels with exercise and a change in diet."

She looked hopeful.

"What do I have to do?"

"The people I'm talking about decided to make changes in their nutrition; they started by quitting sugars and refined products, controlling the daily amount of calories, and increasing the number of veggie portions in their diet. The body doesn't seem to understand 'refined,' it has trouble processing it. It tends to make mistakes when faced with products that are not natural or contain chemicals. These people also walked every day, which makes insulin more efficient, and found stress-reducing activities like meditation and prayer. When their sugar levels went back to normal, their symptoms disappeared and some of them stopped taking medication."

She was really excited now.

"Oh! I wish my sugar were normal so that I'd be allowed to eat the pastries and candies that I like so much!"

As I see it, it's very difficult for us to face our need to heal. This lady's goal was to go back to normal blood sugar levels but she was not aware that pastries and candies caused or at least contributed to her body's imbalance, nor was she looking for a more fulfilling life or willing to learn new life-styles. She seemed to feel she'd been deprived of the goodies she yearned for, felt punished and wanted to be spared the penalty, just so she could go back to her former life-style without realizing that it contributed to her illness.

Reiki is not just a natural healing therapy that involves the *laying on of hands*; it also promotes and facilitates a process of personal transformation.

What could a Reiki practitioner do in her case? Probably help her fathom that her well-being was in her own hands. Help her see that healing is more than the removal of a symptom or the normalization of a lab test. That, as a quest for wholeness, healing involves the amplification of our awareness to the point where we listen to our body expressing its needs and ringing its alarms through symptoms, so that we can respond by becoming responsible for our own wellness and our relationship with the world.

Curing vs. Healing

Medicine focuses most of its efforts and research on palliative and curative treatments.

Curing means removing a symptom. It refers to our physical dimension. A curative treatment gets rid of the symptoms and illness, and a palliative treatment is used when, not being able to achieve the cure, we aim to at least attenuating the symptoms. In the case about the woman mentioned before, the cure or palliative result would be making the blood sugar levels return to *normal* lab levels, which somehow would prevent some of the harmful consequences. This apparent improvement will continue as long as she takes her medications and visits her doctor regularly.

If we suffer from pneumonia, the medical treatment will usually focus on the affected organ, the lungs, and on the germ that caused the disease. Antibiotics will be prescribed to fight the infection and measures will be taken to assist respiration if needed. Intravenous liquids may be administered to prevent increased acidity resulting from breathing difficulties and to replace liquids lost due to fever. Medication may also be needed to decrease the latter. Apparently, all fronts are covered.

Allopathic medicine aims to cure and such cure is focused on fighting disease and symptoms. However, why did pneumonia develop in the first place? Why now? How is the immune system working? Answers to these questions will result in getting rid of the diseases with a better understanding about our body and what we can do to keep it healthy. It will lead also to prevention of new symptoms.

When, instead of pneumonia, we talk about an illness we have classified as mental, something similar will occur.

Depending on how serious it is and how much the ability to function in life is compromised, we go to the psychotherapist in search of "a talking cure" or to the psychiatrist for him to intervene with our biochemistry.

The first option, psychotherapy, offers more chances of facilitating a process that will lead to the expansion of our awareness and to a wiser spirit. Medications, on the other hand, may offer a temporary relief while promoting dependency to chemical solutions and no personal growth.

Healing is not limited to physical processes. It is closer to the concept of enlightenment and spiritual growth that predominates among eastern philosophies presently influencing the West.

Healing is the product of our inner search for lost integrity; the developing and broadening of our awareness that allows us to recognize ourselves as creatures of the universe. It helps us assume responsibility over our body, our actions, our environment, our relationship with others, with ourselves and the world.

The act of healing can be triggered by a physical process. Illness becomes a guide from which we begin our learning and transformational process.

Healing goes beyond removing or attenuating a symptom. First, it involves time and our active and aware participation. Nobody else can live and go through our own processes. These cannot be prescribed from the outside; we are responsible for them. Second, this is a learning process and learning is not just acquiring information or memorizing data, although information is also important in the developing of knowledge. Learning is the process of assuming a new response to a given situation. A learning process involves taking an inventory of our responses, habits and patterns, and getting rid of those that are not optimal from a functional point of view.

To achieve our goals we need to know ourselves, be aware of those beliefs that limit the development of our potentiality, have the willpower to dismiss responses that no longer work, explore possibilities and be free by choosing healthy options. Healing is a transformation process that comprises every level of our existence.

Many paths lead to healing. Each of us will choose whichever attracts us the most or better facilitates the achievement of goals. During that process we will rely on our own strengths (will, discipline, studies); our masters and teachers – temporary or not –; external agents (crystals, essences, colors, Feng Shui, healing therapies such as reiki and psychotherapy); and healing environments in which interpersonal relationships are characterized by respect, validation, mutual support and by how much they contribute to our personal growth.

Healing processes are necessary because we have *torn* our existence: we have separated mind from body, art from science, intelligence from wisdom, feelings from sexuality. We have disconnected ourselves from nature; we have denied our essence. This divorce finds an outlet in the overspecialization within the various realms of knowledge, which in turn deepens such divorce. We live in a world in which we cannot see the forest for the trees.

Healing refers to the need to unify what has been separated; the goal is to achieve wholeness and understand who we are in every single dimension as individuals and collectively. This requires the raising of our awareness, recognizing that we are a spark of the universe and refraining from denying those aspects of ourselves that we reject or find too difficult to accept.

Sometimes I joke about how I used to be a *better person* before becoming a reiki practitioner. After the second level, many aspects of my personality hidden in the darkness, where I could not be aware of their existence, came afloat. I came in contact with feelings I would never have suspected I had, and it was difficult to accept that I had them. The turbulence was disturbing and my self-esteem was notably altered. After a while, I had to come to terms with my human nature. That acceptance allowed me to realize that it takes effort to maintain the awareness of being a soul going through a corporeal experience.

I no longer seek *perfection* as a human being but rather the acknowledgment of my essence. Sometimes I have to prevent myself from becoming too suspicious of my own motives and give myself a break whenever I identify my ego's weaknesses or mistakes because it predisposes me to self-flagellation. I've learned to be compassionate with myself and the immense limitations of my human condition.

Increasing awareness should not lead us to developing feelings of guilt but rather to accepting the infinite potential of our souls and assuming responsibility for our actions and omissions.

In the case of the diabetic woman, she can add meaning to her life if she becomes more aware of her body, the consequences of not taking care of it and the benefits of being healthy and full of energy to enjoy life fully, an enjoyment that is not based just in two minutes of palate pleasure from eating candy.

What is the relationship between this woman and her body? Why would she sacrifice it for a few sweets? She could heal herself. We are all healers – some potentially and some already developed.

The paradox of progress

Even though we now have richer evidence that health depends on us and on the lifestyle we choose, there is still an increasing dependency on external agents such as prescription drugs and surgery when trying to regain our health. This paradox seems to be proportional to the technological improvements in biological science in the last few decades.

I know of people who organize their lives around health services: dentist on Monday, family doctor on Tuesday, orthopedic on Wednesday and so on…

Only fifty years ago, my mom knew exactly what to do when one of us had diarrhea or chicken pox. For her, the Hippocratic principle of food as medicine was sacred. She also understood that, for convalescence to take place, rest and a favorable environment were required.

Nowadays, our wellbeing seems to depend on all sorts of pharmacological solutions. I recently read in a newspaper editorial section that 12 minutes out of each 30-minute block of television news are commercials and most of those promote the use of prescription medications. Ads invite us to favor this pain killer over that one or to suggest to our doctors to prescribe us certain drugs that are presented as a panacea for our ailments – even the ones we may not even suffer from.

Entire lanes in the supermarket offer us pills to treat coughing, pain or indigestion. There are also ointments, vitamins and minerals, and energy bars full of chemicals that replace a meal. People even include these items in the regular grocery shopping list.

Most pharmaceutical and other chemical products perturb those mechanisms that maintain the stability of the internal environment (*homeostasis*). They may, for example, make us stay alert when perhaps what we needed was rest.

Take Excedrin, which is administered for common colds. People get colds when they are exhausted (the immune system weakens with exhaustion) and the body is crying for a little rest. But people are prompted to continue working even when they feel sick, so people suffering from colds take Excedrin with caffeine in the morning and go to work and then take Excedrin with diphenhydramine in the evening, which makes them drowsy, so they can go to sleep.

There is almost no prescription or over the counter drug that does not cause unwanted secondary effects. But of course you can find other prescriptions that will try to counteract such secondary effects: anti-Parkinson for the trembling caused by antipsychotic drugs, antacid to prevent gastritis caused by anti-inflammatory prescriptions, anti-ulcer drugs for those prescribed with cortisol-based medications. As a result we bewilder the body and risk our inner balance.

We continue taking all these products, even when we are being offered better, healthier options, because they seem reliable, researched and... fast! Pharmaceutical companies continue to produce them because our modern world is one of easy and quick solutions and we are in a hurry to get rid of symptoms. We also choose the prescription solutions because we forgot or simply do not know any other paths, or because our health insurance does not pay for alternative modalities.

We cannot underestimate the hope that medicine has offered in such cases for which we previously had no solutions. There are, for instance, antibiotics that have saved many lives and vaccines that have contributed to the prevention of fatal diseases such as small pox. In our present world where there seem to be more chances of suffering more than one accident in a lifetime, surgery offers solutions that were unimaginable a century ago.

Technology allows us to unveil the intimate secrets of our cells and diagnostic improvements – outstanding by the way – mitigate our curiosity, our need to know what, if anything, is wrong with us.

But why are we satisfied with a partial answer? Why do we believe that the key is finding the organ in which the body's dysfunction is manifested? We treat the organ or system that we believe to be ill without first pondering why or how the body's balance has been compromised; before weighing the factors that might have led to such imbalance and that we need to find in order to properly stimulate the body for its recovery.

Generally, after a diagnosis, even if it is a result of exhaustive and sophisticated screenings, we own only a bit of the truth if we leave aside the circumstances that created illness, the context.

The rural doctor, who works in small communities, is familiar with the area's nutritional habits and environmental conditions, and has certain knowledge about her patients' relationships and support networks. He may even have close relationship to the patient's family. This doctor is more likely to integrate such knowledge in his assessment before prescribing or making recommendations to the patient than those doctors who base their knowledge only on sophisticated tests.

Even when it has become commonplace to talk about globalization in both the economic and political fields, and the word *holistic* is becoming part of the medical jargon, we still live in a world of predominantly disjointed views.

A client came to me with an excruciating pain in her shoulder. She wanted explanations, a diagnosis and a quick fix. As a reiki practitioner, I provide none of these. She lay on my massage table and my hands went directly to her shoulder, one hand above, one hand below. While I was sitting beside her, feeling what we call *universal energy* flow through my hands, a sudden thought came to me. I had learned the anatomy of the shoulder; kinesiology had instructed me how muscles interact to produce movement; physiology had provided me with explanations of how muscles contract and about the biochemical and neurological aspects of it; I had studied pathology and thus knew how muscles, ligaments and bones are affected by disease and injury.

However, sitting there, I remembered how frustrated I felt on occasion about the incapacity of my own doctor to understand what was going on in my body. I became aware that, although I could provide some guessing and try to answer my client's questions, I actually knew nothing about this woman and her pain and, what was most interesting, it really didn't matter.

She and her body were the real experts. It was her body that knew how to repair the injured tissues and what substances to release in order to alleviate the pain. Reiki was just an enhancer, a tool to support the *inner healer*.

Twenty minutes after I laid my hands on the shoulder, the client cautiously moved it. The pain was almost completely gone. She could still feel a limitation to movement and I encouraged her to respect her body, which was calling for a pause while completing the repairing task. She continued coming for sessions and had no need to take analgesics, which usually interrupt the communication among organs and might delay the healing process.

Because I know that Reiki doesn't replace medical attention, I recommended that she visited her doctor. She went to her primary-care physician, who at first was reluctant to order an MRI. At her insistence that she had the right to further examination, she was referred to an orthopedist. Because he could find nothing in the MRI of the shoulder, he ordered an MRI of the neck. As the test showed no pathology, with only minor changes due to the normal aging process, he referred her to a neurologist. She explained to this specialist that she was also having some burning sensation on the left side of the chest, mostly under her breast.

When the neurologist found nothing that could explain the burning sensation from his perspective, he decided to send her to a cardiologist, who couldn't find any pathology.

She was also referred to an Ob-Gyn to rule out a breast problem. She refused to have a mammography and it took a while to convince her doctor that she preferred a thermography for a diagnosis. She had been reading about both and decided that she didn't want her body to be exposed to a high dose of x-rays and that thermography could probably offer an even earlier detection of a mass, if that was what she had. Finally, a gastroenterologist saw her and recommended some medication for reflux, not because he had found anything wrong with her, but "just in case." The medication made her a little drowsy, so she discontinued it within days.

I could say that this four-month search for an answer to her "what's-wrong-with-my-body, with me?" question served the purpose of calming and finally convincing her that she was actually healthy. But the cost! In terms of dollars and in terms of time! Fortunately for her, most of the expenses were paid by managed care, except for the thermography. But what happens in the cases where the patients cannot afford all this medical visits and exams?

This story also reminded me of the small villages where I many times worked in my country. In these far away places with few medical resources, where patients don't have the money to go to the city for lab tests or any other kind of exams, rural doctors need to sharpen their diagnostics skills and work very closely with the patient from an integral perspective. They have to provide nutritional advice and probably make recommendations that would normally fall within the competence of other specialists.

It's frustrating sometimes, but it also forces the doctor to be in tune with her own intuition ("clinical eye," we call it) to do the best possible for the patient.

Trusting the inner intelligence of the body has completely changed the way I understand my role as a professional, has humbled me and has made me realize that I don't fix people. No matter what the condition is, the real expert is the body.

Molecules as messengers

The reason medications work in the body is that there is a similarity between the pharmaceutical administered and a substance that the body produces naturally. The best example is morphine. It works because we produce our own opiates (called endorphins) and thus our bodies have receptors to which the endorphins would usually anchor. Morphine may be administered when, for some reason, endorphins are in short supply or the body needs a larger amount of them, like after a surgery or trauma.

The drug molecules *marry* to the body's natural receptors and "clog" them. The problem is the receptors (which are not exclusive but shared with other substances) keep busy and inform the body that there is no longer a need to continue producing the natural substance, which is now provided by an external source. This in turn compromises the communication between organs, which is mediated by the natural molecules. And this is what explains the secondary effects of medication. We will come back to this later on.

In the woman's case discussed in the previous section, her ailments began a few years back after she went through a hysterectomy (removal of her uterus) due to fibroids and her doctor decided to extirpate the ovaries. Ever since, she had been on hormone replacement therapy. My hypothesis is that these hormones disturbed the inner communication and that many of her symptoms were related to this disturbance, including depression.

Elena[9], a friend, returns from her doctor. "They diagnosed me with colitis and prescribed me a diet," she says. The doctor's recommendations, she explains, have the purpose of avoiding those foods that may irritate the colon.

However, as we know, most foods have already been chewed, mashed, mixed with enzymes and absorbed by the time they get to the colon. Only residues of this process make it to the larger intestine. How can, then, a diet make her colon's condition change?

The medical argument is that some foods (e.g. coffee and colas) interfere with the bowel movements or with absorption. Also, that there is a need to increase the intake of fiber in order to prevent constipation because if the transit of residues is delayed, it may cause the products of fermentation, alcohols and gases to irritate the bowels.

But, what is colitis? According to our multidimensional and holistic perspective, it is – in the physical dimension – inflammation of the membrane lining the colon usually associated to an excess of ferments, parasites and/or bacteria. When we mix fruits and vegetables, or sweets and salty foods, and whenever we abuse sugary, protein or fatty foods (e.g. meats and dairy), digestion becomes difficult. On the other hand, the indiscriminate consumption of antibiotics and processed foods changes the intestinal flora and modifies the production of digestive enzymes, which in turn makes things worse.

Most of you have experienced how, in a very stressful situation, there may be a sudden urgency to visit the toilet, or how this same visit may become challenging if the person cannot relax. This shows a connection between the digestive process and emotions. The latter

---

[9] Name has been changed.

could interfere in the secretion of needed digestive enzymes or in the peristaltic movements.

At first, the body's response to an inadequate diet may only be slight pain or diarrhea. However, if nutrition remains the same and the emotional turmoil continues, the inflammation will become chronic.

A chronic process will exhaust the body's resources. At first, the irritant will cause the immune system to concentrate on the colon, the endocrine system will promote the production of more enzymes in order to accelerate fermentation, and the nervous system will make an effort to rush peristalsis so the body gets rid of the nuisance. Intestinal mucous cells secrete liquids so that feces become blander and easier to excrete.

The entire body becomes alert in order to get rid of whatever is cause of inflammation. If the situation takes longer, though, the *inner healer* becomes exhausted and the communication among systems becomes dysfunctional. In addition, the mind's mantra "I have a problem," increases the levels of stress further compromising the immune system functions. The resulting imbalance leads to inadequate responses and then, most foods will cause additional fermentation.

Healing then would not involve worrying about ingesting foods that do not irritate the colon, but rather following a diet that contributes to restoring the ideal inner conditions.

As you see, the formula is the same for all conditions: we need to avoid processed and refined foods, certain fats (such as animal fats, corn or sunflower oils – rich in pro-inflammatory omega-6 fatty acids). Also, we need to increase intake of vegetables and fruits that are rich in fiber, vitamins and minerals, and drink yogurts containing *probiotics* that restore the intestinal flora (see nutritional recommendations on page 190).

It is worth noting that when a person is suffering from chronic diarrhea, he might need to avoid raw food until balance is restored. He might also need to be treated with reiki, acupuncture or any other energy field therapy that can restore the flow of energy in the body.

It is interesting to note how in the most-respected North American scientific publications authors report having reached similar conclusions about the variables that influence the occurrence and outcome of different diseases but without integrating the information at the time of interpreting findings or drawing conclusions. Let me explain.

Reports show that cardiovascular disease, cancer and diabetes are the leading causes of disease and death in the United States. Arthritis is also a condition of frequent occurrence.

Many scientific articles point towards the relationship between rest and diabetes, rest and arthritis, rest and cardiovascular disease, rest and cancer. Shouldn't we just construe that rest promotes health?

There are also many who conclude that nutrition is a definite factor in preventing those diseases. Studies proving the benefits of physical activity in those conditions have also been published. In spite of the popularity of new diets on the market, there is consensus about the existence of some foods that promote the proper function of the body's systems – especially of those that are in charge of the communication between organs– and other foods that interfere with such process. Conclusions, though, are still presented only partially.

Let's integrate. Let's say that there are three basic pillars to prop health up: nutrition, physical activity and adequate stress management, which include adequate rest.

Each chemical product that we introduce in our bodies (and food that is neither natural nor certified as organic does contain chemical products) occupies the receptors for the messages that the body dispatches to go determine the functioning of its parts, promote tissue regeneration and repair, or offset imbalances. When these receptors are busy, messengers cannot deliver their message. It is like when our electronic inbox is full of junk mail and our close friends can no longer reach us.

Educational campaigns promoted by health authorities have proved to have positive results, as seen in the national decrease in smoking and a certain obsession for no-cholesterol foods.

During the many years that have passed since I started medical school, I have seen a shift in the focus of health campaigns, which shows an increased understanding and a forward movement towards what's natural. For instance, when I still was at medical school, I observed the shift from promoting baby formula, vitamin-enriched chocolate beverages and powder cereals for babies, towards campaigns that rather focused on inviting mothers, especially those of limited means, to return to the benefits, economic advantages and wonders of breast milk. During this time, I also observed how the medical establishment abandoned many ideas, some of which had been already been presented as absolute truths.

Hormone therapies, for one, were defended passionately and indiscriminately offered to menopausal women. Recent studies linking hormone replacement therapies to breast cancer, however, have led to doctors being much more cautious when suggesting such treatments. Also, eating eggs, which had almost become a deadly sin, is recommended (with some restrictions) nowadays due to, among other things, its lutein content that helps prevent blindness. There is increase evidence that medication to help decrease cholesterol also produce dangerous secondary effects, contributing little to the person's recovery. In addition, it seems that cardiovascular disease is being caused by an inflammatory process and not by an excess in cholesterol levels after all.

The acknowledgement of how important nutrition is has led to numerous studies, publications and diets. Nowadays, though, the ghost of obesity wanders around Europe and North America; this being even more the case for those immigrants from third world countries who get fascinated with fast food and have little access to educational campaigns.

Simultaneous with television advertising campaigns – which permanently stimulate the consumption of all kind of delightful but nutritionally empty treats that are high in calories and saturated fats – Hollywood has generated an anxiety to achieve an almost anorexic body in order to follow standards that are not, and should not be, the average population's weight and size[10].

Overweight and eating disorders are not the only result from these phenomena. Diabetes, hypertension and cardiovascular disease risks have increased in the United States and Europe among children and teenagers. Problems in overweight individuals related to low self-esteem and social performance, as well as suicides related to difficulty accepting self and being accepted by others, have become increasingly frequent. On top of this, obese people usually have lower incomes than the rest of the population[11].

The social cost of obesity-related diseases has proved some cultural traditions accurate. For instance, in several Latin American countries, diet is based on the perfect combination of proteins, such as beans with cereals like corn or rice and vegetables. Many advantages have also been found in the Mediterranean diet, and Asian cultures have been eating the recommended daily servings of vegetables for ages.

These nutritional customs vary according to the degree of urbanization that has caused a shift from consuming fresh, local

---

[10] As I write, news report that soon it will be a crime to glamorize the ultra-thin in France.
[11] Jonah Bloom in *Junk science as much a part of 'fat epidemic' as junk food*, published in *Advertising Age*, 76, (24, Jan., 2005).

produce to ultra-processed, packed or canned foods containing sugars in a variety of delicious forms, very high in calories, preservatives and colorants. This nutritional change perturbs the body's ability to maintain its inner balance.

Physical activity decreases as life becomes more urban and domestic chores require less effort as well, and we progressively become more and more sedentary. According to the American Pediatric Academy and the American Medical Association, there is a directly proportional relationship between the number of hours a person sits in front of the television set or computer and excess weight.

As industrialization grows and people choose to live *American dreams* in overpopulated places, not only do we lose our nutritional traditions but we also contaminate the air, the water and the food.

The average American lives isolated, depressed and in a hurry. Children no longer play in parks and gardens. Instead, they concentrate in front of some type of screen: television, computer or *Gameboy*. People move around in little metallic boxes that travel through frantic highway traffic. They barely have time to eat their usually unhealthy *fast food*. They even give up moments that could be of contemplation, private moments, to instead be hooked to cell phones or computers in an effort to protect the illusion of being connected with clients, bosses, friends, family and the world, yet at the same time not knowing who their neighbors are.

In such a world, true needs are never satisfied. In the movie *Roger Dodger,* by director Dylan Kidd, the main character works at an advertising agency and shows us Pavlov's philosophy behind marketing. They make us see our (real of fictitious) lacks and unhappiness, conditioning us to new needs so they can sell us more stuff as a panacea. To sum it up, the modern lifestyle prevents us from becoming aware of who we are; we seldom have free time and we do not take care of the planet or ourselves. We dwell in a world of illusion, appearances, thoughts, images and words. We are split, broken.

Once the connection between our body and its true needs and priorities is lost, we tend to ignore its alarm signals and our heart and we also lose our relationship with the Universe to which we belong. The self-assurance with which we contaminate the world gives account of this as do the many ways we move around and relate to others. The result of these new lifestyles is the weakening of our *inner healer,* which can no longer keep the inner balance, which in turn results in symptoms and illnesses manifested in any and all of our body dimensions.

36

I have also wondered what makes us lose balance and get depressed. The United States proudly claims to have the highest standard of living in the world. It is, for many, the land of opportunity, democracy and freedom. It is presented as the model to which other countries should shape their own.

Shouldn't it be then the land of happiness? Well, according to statistics, that is certainly not so.

Suicide rates have increased 300 percent since 1950 and nearly 2,000 teenagers attempt suicide each day. In the United States, more youngsters commit suicide than in any other industrialized country. Bottled anger has led to sacrificing lives in deadly shooting rampages.

Depression is only good news for pharmaceutical companies that continue to increase sales of antidepressants. Increasingly, Americans are medicated for all types of depression, however mild, even if it's just the normal stage of a mourning process after a significant loss. Doctors argue that depression is a chemical imbalance that should be corrected with pharmaceuticals.

However, even though doctors today are better at identifying depression and labs have developed new families of more effective antidepressant drugs, statistics show no reduction of depression in the population. In a recent *National Comorbidity Survey*, about 14 million Americans reported experiencing a serious depressive episode in the previous year, and 35 million reported at least one episode of serious depression over their lifetime.

On the health of our planet

This patient had walked for many hours. She was coming from far away, from the mountains, for a consultation. She was pregnant; she was 37, and she was single. Because her parents would have disapproved having a baby out of wedlock, she waited until both her parents died and then looked for a handsome, healthy guy and asked him to plant the seed.

She came to the medical center asking for nutritional advice. She wanted to know what a proper diet for a pregnant woman was because she wanted to make sure her child would be born in good health. Although she was not an educated person, she knew that nutrition was paramount. She had heard on the radio that she should take some vitamins and supplements and wanted me to give her a "good" prescription. She talked to me in an intimate way, like we were old female friends, and her voice was like a prayer.

I looked at her, feeling a bit intimidated, thinking that she was the expert regarding her body, and that she could listen to her body and know her needs, but had somehow lost access to her intuition and knowledge. I wondered how I was going to tell my patient that our expensive vitamin pills had less potency than the fruit she could directly harvest.

As a doctor, I chose community service, and thus instead of staying in the big cities where I could have all the commodities, I looked to work with small communities throughout Colombia. In my many years of practice, I witnessed the negative connotations of consumerism, presented as progress, on people living in the countryside.

Transistorized technology had hypnotized them, transforming their nutritional habits and their traditional farming practices. Now, they prefer white rice over whole brown rice, even when it comes from their own crops. And they favor refined sugar over traditional molasses that they produced in rudimentary cane mills.

I've seen peasants selling mangos from the trees in their orchards and eggs from their own free-roaming hens to buy canned food filled with preservatives. They had heard it was nutritious. Ads are so impressive. Hollywood stars that advertise products look so happy and healthy. Where would they read the truth about bulimia and anorexia and drug abuse?

As a doctor, I observed many different nutritional problems. From the malnourishment of the gelid moorlands on top of the Andean mountains, where people have diets rich in carbohydrates and poor in proteins, to the very poor in the hot, humid valleys, where food is lacking and parasites suck blood until they cause anemia, killing their victims.

I've seen malnourishment in the rich guided by whimsical appetites and not by their nutritional need and also the middle class women who want to look glamorous fall sick. (By the way, obesity has been officially classified as malnourishment).

But there were a few places where I lived where malnourishment was not a big issue among the population of peasants who owned small farms and practiced agriculture in a traditional way; whole regions where people owned small self-sufficient farms.

With everlasting-spring kind of weather, the land in these regions renders an incredible variety of produce: sugar cane, fruits, coffee, plantains and yucca. In these places, maize and beans grow embracing each other, as if showing us they should be eaten together for full nutritional value.

Most farmers in these regions have a vegetable garden and their own livestock: at least one cow to provide milk, pigs that are sacrificed during important celebrations and hens that provide eggs.

Peasants go to the market and sell some of the eggs and fruits to buy meat and salt to provide a provision of dry-cured meat (jerked beef) for a few weeks. Everything they eat is fresh and natural, and they don't have the need for a refrigerator. These peasants do not accumulate tons of foods in fear of scarcity and they don't generate waste like we do, neither do they pollute the waters they drink the way we do. They are already reducing, reusing and recycling, and their diet tends to be balanced.

It doesn't mean that these peasants live an ideal life, not by Western standards anyhow. Their access to health and education is limited. And, unfortunately, current school curricula do not help them keep the wisdom that they have passed from generation to generation for centuries. Education generates consumerism instead.

I used to view their lifestyles as a sign of backwardness, but now when I look at the future and understand better how human activity has caused global warming and is threatening life on the whole planet, I think we have much to learn from those we have called primitive and underdeveloped.

In developing countries, and especially since the obesity epidemic was publicly acknowledged as an increasing risk for the health of millions of people, we're trying to reverse the effects of our loss of good nutritional habits. However, I see that even if people really strive to keep a balanced and organic diet, it is very difficult to escape the siege of junk food that is offered in banks, a party or the office.

If somebody, resisting the temptations of consumerism, keeps her life simple, we call that person "tight," stingy, rigid.

In the history of humanity many empires have raised and then sunk because of gluttony (eating more than needed), greed (seeking to have more than needed), lust (life guided by the pleasure principle) and arrogance (imposing our ways to others, mindlessly exhausting the planet and thinking we're doing well). We need to learn from those mistakes.

Since the beginning of extensive agriculture 10,000 years ago, men have sought to expand their domains with no vision of the future and with no care for fellow humans. This process has been responsible for the devastation of entire regions, creating extreme inequity.

In a third-world country, peasants who don't own land colonize the forest or the jungle and establish their quarters there, where life conditions are extremely difficult. In those isolated places, challenges invite people to befriend in order to grant survival. United they can produce enough quantities; they can commercialize surplus, pay for what they do not produce and ship supplies at a lower cost.

Among the harsh realities I had to witness was the lack of options people in third-world countries have to maintain healthy life styles. There are peasants who own no land; people earning wages so low that they cannot pay for the most essential goods; people living in far away places where sanitary conditions are at fault, where supplies are insufficient or extremely expensive. In many cases, they ignore the healthiest choice because they have lost natural instinct due to modernization, and this lack has not been replaced by instruction and education.

I have listened to those who defend, and I have defended for years, industrialization as a synonym for progress. But presently I think progress depends on the use we make of technology and science. Technological advances can spare humans much suffering and can mean the betterment of our life conditions. But humanity is complex and, in today's world, industrialization's main features are merchandise and profit, with underlying destructive greed. Greed many times leads to violence, expansion and power struggles. Technology must not be used under a paradigm that assumes that we humans own the Earth and thus we can exhaust it, without vision for the future.

In many countries, even if you have the best of the intentions to have good nutrition and you pledge to consume only products that have not been chemically processed and produce from orchards that have not used chemical fertilizers; you avoid refined products and drink only milk from cows that have not been treated with hormones or antibiotics … even then, people depend on the cost and availability of these products that, at least for the time being, are costly and scarce in third-world cities.

I have seen affluent people arriving at crossroads where they've had to question their nutritional habits and initiate a process of change. Frequently they just scratch the surface. Underneath the changes they made, there was not enough will or determination so they didn't bring about any real benefits. They continued cooking with no love, storing food in the fridge until it got rotten and eating in a rush.

Thanksgiving rituals are lost and we fail to remember how fortunate we are that we can bring bread to the table. In Alfonso Arua's movie *Like Water for Chocolate,* the Mexican director presents a melodramatic plot where Tita, whose love has married her sister, discovers that her feelings surface through her cooking. She takes revenge by manipulating the emotions of all who eat her food. The film is based on rooted beliefs present in many cultures: that what we feel influences the taste and good of what we cook.

Nowadays we know that endorphin secretion increases during pleasant activities. Guess why we like junk food… we are rewarded with it. Pleasure makes organs communicate in a more efficient way. That's why we should take meal breaks so that we can eat mindfully and nourish ourselves properly. There is no such thing as *healthy fast food.*

Regaining wholeness includes becoming conscious of the possibility of oneness with all that surrounds us and turning our sight towards the earth that nourishes us. If we want the planet to continue feeding us and generations to come, we must respect it, care for it, and honor it. And for that, we need to build a different kind of society.

Sociologist and author Daniel Quinn brings the issue in his books *Ishmael* and the *Story of B.* In Ishmael, the main character, a gorilla, discusses how humans moved away from living according to the *laws of life*, the same laws followed by all other living beings on the planet. The main difference between us and other creatures is that they belong to the world, while we believe that the world belongs to us, which explains our greedy behavior.

In South Florida – my home town at the moment in which I write these lines – most of my neighbors have sprinkler systems to keep their lawns looking like a uniform green blanket during dry season – November through March – when because of lack of rain, the grass dries out. I learned that dried grass serves as a protective coating against the soil's excessive dehydration. Also it will rot, and make humus, acting as a fertilizer, and little flowers and shrubs will start growing through the straw-colored grass. White daisies and little violets as well as resistant creepers start climbing up the rough pine cortex. Gardens become the residence of doves and butterflies. It is a different flora and fauna from the one that is seen during the rainy "summer." It is the life cycle whose secrets we still do not completely unveil.

My neighbor used to spray to kill weeds. I used to joke with him asking what was wrong with those plants. He smiled back and scratched his head. After many years performing the same routine, he had forgotten the reason! Why do we really get rid of weeds? Because farmers taught us so. For them, weeds result in a waste of money when they spread around their crops. Wise agriculturists rotate their crops and understand that every single thing created by nature serves a function.

Why do we insist on violating the Earth's cycles, forcing its productivity and filling it with indestructible wastes? Those transformation processes of the Earth, extensive agriculture and technology have not made the hunger in the world cease. Abundance of the few does not guarantee the affluence of the many, and extensive agriculture has made deserts of places on the Earth that currently lack drinking water, and thousands of children are still dying of starvation.

Two billion people in the world – one third of its total population – do not have access to drinking water. Behind our mindless expansion lies our greed, our haste to possess more than we need, the ambition to become proprietors, our fear of not being able to provide enough for survival.

It is understandable that humanity lives frightened about the possibility of repeating starvation times due to drought. We have come a long way in learning how to store and maintain food in planning for potential harsh times; however, we are milking the planet dry until turning it into a sterile place unable to maintain life. We are killing the hen that lays the golden eggs because we are not being rational and sensible enough.

A while ago, when walking around the neighborhood in the morning, I felt the ground move, trembling in response to a nearby bulldozer. A couple of acres of raw forest were being flattened down to build a house model and a grass garden in its place; they dug the ground to build the foundation. Doves, squirrels and shrubs were gone. The whole picture was devastating and I felt then and there that if we cause devastation, no matter what we are doing and no matter how we justify it, we cannot call this development and progress.

If what defeats our *inner healer* is the interference with the communication between organs of the body, which in turn prevents the successful performance of its homeostatic functions, this also applies to planetary health. The Earth's ability to maintain stability is affected by the way we treat it. Its changes are so slow that we live without concerns about how we are affecting it. But these changes are occurring faster every time, as we have witnessed in the unusual climate phenomena[12] whose explanation is found by many scientists in the greenhouse effect.

The societies we have created deteriorate the quality of our waters, food and air. Our relationship with the planet, the roles we play in the society, individualism and the quality of our relationships with others are the origin of most of our problems. There has been much talk – but not enough practice – of loving our neighbors. Our lives are full of stress, fears and frustrations.

Quinn, mentioned earlier, divides humanity in two main groups: the *takers* and the *leavers*. The first ones invade, possess, transform, violate. The second ones respect, observe the *laws of life* and live accordingly. We call some of the latter, primitive communities; few of these remain and their legacy rarely reaches new generations.

According to their beliefs, the water from Lake Okeechobee, the second largest of the United States, was offered to the Miccosukee tribes by the "Breath Giver." These Native Americans, who migrated to the Florida Peninsula 200 years ago, lived from hunting and fishing and some of them cultivated corn.

The integrity of the 5,000-years-old Everglades ecosystem, the large and slow- moving river that flows from Lake Okeechobee in the heart of Florida to the Gulf of Mexico, started being threatened by the arrival of Man who built dikes to create real estate properties and later to prevent floods, and provide newly populated areas with drinking water.

In 1909, the canal that links Okeechobee and Miami was completed. Later on, the Tamiami Trail – a rural highway that runs from Miami on the East to Tampa on the West – crossed the State through the Everglades, cutting off the water's natural flow. The trail crossed the most beautiful area, known as *Big Cypress*, a deep forest of about one million acres where little islands can still be found, where orchids hang from trees and panthers hunt otters, deer and porcupines.

The Lake began to get all the phosphoric waste from the agricultural industry, especially from sugar cane. Nowadays, its waters are so contaminated with pesticides and mercury buildup from former

---

[12] For instance, a record number of six hurricanes caused devastation around Gulf of Mexico during 2004 and 2005, and a Tsunami in the Indian Ocean affected all adjacent countries by the end of 2004.

waste and battery manufacturers that a great deal of game fish cannot be eaten, most vegetation is ruined and hunting has decreased.

Thirty years ago, when Miccosukee tribal chief Buffalo Tiger endorsed the campaign to make his tribe into a nation, he did not imagine that, even though their right to the land would eventually be conceded, they would no longer be able to return to their former relationship with such territory.

Nowadays, natives have become tourist guides taking tourists for a ride in airboats that with their huge propellers open paths without even touching the waters. Those who are not natives proudly talk of the *civilization* process of the *Indians* who have adapted and incorporated to the flow of *progress*: they own casinos, air conditioning in their houses and SUVs, which incidentally are the vehicles that consume the most gas and that contaminate the most.

Adapting, mitigating and awareness necessary to deterring climate change

Many people visit and move to Florida every year, lured by the pristine waters of the gulf of Mexico or the white sands of the Atlantic Ocean on the East Coast of the peninsula. The state offers opportunities for recreational fishing, and promises a variety of entertaining activities.

Regrettably, the seductive quality of Florida's getaways is at risk because of climate change.

Global warming is no longer just a scary headline on a newspaper. A consensus already exists among the scientific community: global warming is already affecting the globe and human activity is changing the atmosphere of The Earth at a rapid pace.

The Intergovernmental Panel for Climate Change (IPCC), which shared the Nobel Peace Prize in 2007 with U.S. ex vice president Al Gore, released that year an alarming and convincing report about climate change that has created a sense of urgency to take action in the United States.

The IPCC was set up by the World Meteorological Organization and by the United Nations Environment Program. It is made up of a group of experts who carry out comprehensive assessments of all aspects of climate change. They have gathered the most talented contributors from around the globe and follow a rigorous process of peer reviews. Their reports get the final acceptance by governments through a process of consensus.

In their last report (AC4) released on November, 2007, the IPCC presented detailed evidence of global warming. The average air and ocean temperatures have increased; there is widespread melting of snow and glaciers and ice and average sea levels are rising.

IPCC reported that efforts made by the governments who signed the Kyoto protocol 11 years ago seem weak and green house emissions continue to increase, with a considerable jump of 70 percent between 1970 and 2004. "The atmospheric concentrations of carbon dioxide and methane in 2005 exceeded by far the natural range over the last 650,000 years," according to the IPCC report.

Does it mean that the dreaded future predicted by visionaries in the 1980s has arrived? Should we prepare for a future of droughts and flooding, catastrophic weather events, unsupportable temperatures, melting glaciers, water-stressed populations?

As the global average surface temperature raises, so does the global average sea level and at the same time the Northern snow cover decreases. In early 2008 rock analysts found that West Antarctic Glaciers are melting at 20 times the former rate.

The cycles of glaciations and thaw repeat every 40,000 to 100,000 years, but the IPCC attributes the current Earth's unprecedented warming rate to human activity.

Close to 40 percent of the world's population currently lives within 60 miles of the coastline, and the U.S population is rapidly growing in the 50-mile area close to the seashore. Not only this population will be at risk if sea levels rise as it has been predicted, but also the environment is increasingly affected by this urban growth.

In Southwest Florida we see the phenomena of red tides and drought signaling that climate change is occurring. Changes in mangrove population have also been observed.

Action needed now

The White House and federal agencies officially acknowledged in 2007 that global warming was happening. In the summer of the same year, Florida's Governor Charlie Crist also recognized the issue, forming an action team with experts from both the private and the public sectors and he later signed a set of groundbreaking executive orders to address climate change.

Even though the United States has not agreed at the federal level to accept the gas emission caps, and the Congress has not passed one single bill to cap and reduce America's global warming pollution, several states and counties of the union have adopted regulations.

Florida governor's orders demonstrate his commitment to achieve a reduction of gas emissions by increasing energy efficiency and exploring renewable energy sources such as wind and solar technologies. Crist's guidelines match standards already set by the state of California and are also at par with Kyoto's recommended standards. Unfortunately, in December 2007, the EPA denied California and 17 other states the right to apply regulations to reduce emissions from new automobiles. It's still unsure if the Environmental Protection Agency will eventually allow the states to set up higher than federal standards.

The Kyoto protocol is an agreement made under the United Nations Framework Convention on Climate Change, which the United States signed at its inception but, at the time I'm writing this, still has not ratified. The countries that have ratified the protocol made a commitment to reduce carbon dioxide emissions and other greenhouse gases.

By the end of 2007, Kazakhstan and the United States were the only signatory nations that had not ratified the act. The protocol would have required that the United States reduced emissions by 7 percent below the 1990 levels, by 2012.

Last December, the Kyoto signatories met for the 13th session of the Conference of the Parties to the United Nations Framework Convention on Climate Change (COP-13) in Bali, Indonesia. The Bush administration didn't concur with guidelines for reducing greenhouse gas emissions during the convention but by the end of the meetings, signed an agreement for further negotiations over the next two years.

One of the main concerns that prevents business leaders from acting and stopping pollution of The Earth is their fear of profit loss. Do humans have to choose between economy and environment? Can the problem be actually fixed?

Environmentalists seem to agree that what is really costly is inaction.

Automakers like General Motors and Ford that have not yet shifted to efficient cars are seeing their revenues dropping to the point of billionaire losses, even though they have the technology to shift. Only recently GM pledged to sell thousands of Chevrolet Volt, a zero-carbon emitting electric/gas/biofuel compact car with a range of about 30 miles. And Ford has launched its Eco-boost engine that can be adapted to different vehicles, from small compacts to large trucks, promising to increase fuel economy up to 20 percent with reduced emissions.

"Leading companies find gold in green," said Don Miller, a speaker trained by ex vice president and Nobel Prize Al Gore. Miller is optimistic. Can we fix the problem? Yes, he said, certain that the technology needed to reverse the climate change trend is already available in the United States.

"If we use the right methods we can reduce emissions below the 1979 mark by 2050," he said at a local 2008 Focus the Nation event in Collier County, even though some scientists say that won't be fast enough.

Planetary health is as important as our personal health and our survival depends on us having it clear. We are co-responsible with the universe. And to understand this we need a paradigmatic shift also.

From the epic dreams that we flagged in the 1960s and 1970s; from our pledge to transform the world with our sacrifice and effort into a better place to live, we are moving towards a more modest task where we take over the commitment of becoming more aware and better human beings, educating with our example, our words, our sense of community and our civic endeavors.

Through history, individual and group lifestyles evolve. Some people and uses succumb to devastation caused by others. Also, some cultures melt into one, sacrificing some of the wisdom stocked up for centuries. It is interesting to note that the governments and multibillion corporations' push for globalization coexist with a retro tendency that invites us to return to our origins. Some people champion development understood from a new perspective, to protect the species that industrialization and urbanization have placed at risk of extinction.

In 1978, the World Health Organization conference on primary care in Alma Atta launched a declaration proposing a *Health-for-every-one-in-the-year-2000* goal. The importance of this declaration resided in the recognition of the fundamental role that governments play in promoting health and preventing illness.

The meeting steered away from health policies focused on remedial strategies, which had been government's focal point until then. For the third world countries it meant important changes. More than building more hospitals, it became paramount to train rural health promoters, develop vaccination campaigns, and look for solutions for the health problems due to a lack of sanitary infrastructure.

The need to develop centralized health systems that emphasized prevention was sustained on statistics. In the so called *under-developed countries*, three quarters of the people requesting medical attention suffered from conditions that were either self limiting (like chicken pox or colds) or were the result of poor sanitary conditions and unhealthy lifestyles promoted by advertisement.

We already turned the century and we have not achieved the health goal that the WHO proposed in the 1970s. Preventable illnesses and deaths continue to be at the top of the list and the trend is not towards global health care but towards privatization.

On January, 2008, research supported by the *Commonwealth Fund* and published in that month's issue of *Health Affairs* revealed that the United States was last among 19 developed countries when it came to preventable deaths. Analysts attributed it to the lack of access to health care by almost half of the nation's population. Also, to the fact that private companies focus solely on the remedial.

As the result of privatization, many governments have relinquished their obligation to educate and provide health services to their people; costs have risen and the quality of services has dropped at the same time that access to health continues to be restricted to the few who can pay expensive insurance premiums. In the United States, only 44 percent of the population is insured.

Privatization has generated a counterpart. People have turned to different forms of self-care, such as acupuncture, chiropractic care, QiGong (or Chi Kung) and reiki. At least two thirds of the U.S. population is using alternative and complementary medicines, which are kinder, less invasive and less expensive.

In the past 30 or 40 years, the world has witnessed a movement towards medical practices that emphasize the search for optimal functioning and are not just focused on eliminating symptoms. The approach tends to be preventive and based on self-care practices.

The logical conclusion is that we should be shifting to healthier lifestyles, minimizing the use of pharmaceuticals, toxic substances and preservatives and moving towards the promotion of healthy behaviors.

In a world ruled by corporate interests, this shift might be making alternative health become a trillionaire business, and we are being bombarded with propaganda about products that, offered as panaceas, perpetuate the idea that health doesn't come from within and that we could not be healthy without the help of an expert or of some kind of external product.

Noxious agents – sickening stressors

All over the world, a popular medicine that is not registered anywhere, is practiced. Many people seem to give certain sanctity to medical knowledge that echoes the mysteries of religion.

Maybe as a consequence of privatization, health care has become so unavailable cost-wise that people are looking for alternative solutions when they fall ill. This popular medicine, practiced by all those who for some reason have become familiarized with health matters, is simple.

In the times of the great Moliere, many people suffered from *colique de Miserere* (appendicitis); a century ago all illnesses were deemed contagious and a couple of decades ago, every difficult diagnose was considered a viral infection. Now there is a new diagnosis that seems to fit all.

The neighbor complains of headache, a colleague says he has chest pains, a nephew suffers from a urinary tract infection and all claim in unison: "It's caused by stress!" It's known that the neighbor's husband is cheating on her – what a stress! His co-worker is late for a deadline – he is so stressed. The nephew is failing her college semester –stress is making her ill. In this case, they are all right.

Stress is the result of a change in the environment perceived by the body or mind as a challenge, a threat or a factor that can throw the body out of balance.

The body, however, is born equipped with everything necessary to adapt, because stress is part of the normal process of interaction between human beings and their environment; it is a normal part of life. What is new is chronic stress, characteristic of modern life. Never before have humans had to deal with so many varied and intense stressful factors, with so few restorative pauses and compensating elements.

All stimuli that challenge the body cause stress. If intense, it causes harm, pain and *dis-ease*. Even when it is not excessive, but rather repetitive, it causes an imbalance that we know as illness.

The increasing contamination of our environment and food causes new kinds of stress with harmful effects on the body that are explained mostly by the increase in free radicals. Oxidation processes that comprise the transition of a couple of electrons from one atom to another are normal. On some occasions, a molecule with a weak link is broken which leads to an incomplete number of electrons in each part of the split molecule and these are called free radicals. Even as a normal part of its functioning, the immune system produces some free radicals to neutralize viruses and bacteria.

Free radicals are highly reactive chemicals that attack molecules crucial for cell function by capturing electrons and thus modifying chemical structures. They affect metabolism, hormonal activity, synthesis of genetic material and cell behavior. Free radicals disrupt patterns of electromagnetic energy in muscles and destroy the protective fats in the cell membrane, leading to fluid retention and accelerating the aging process. Many degenerative diseases such as Alzheimer's are now linked to an excess of free radicals.

Although free radicals are a normal byproduct of cellular metabolism, an excess of such ions in response to electromagnetic fields or toxic chemicals might render the normal antioxidant defense system of the body insufficient and incapable of preventing the occurrence of disease.

Free radicals behave as *spinsters* whose only purpose in life is coupling up. The same way society produces a number of *spinsters*, free radical formation is a natural process that happens inside our bodies and that also occurs in foods when processed, when fried, roasted, cured by freezing or when irradiated.

Free radicals are very reactive and unstable and they try to steal other people's couples (electrons) in order to form more stable compounds. A chain reaction is then created to form new free radicals. Usually, the body can cope but if there are not enough available antioxidants or if free radical generation is excessive, cells will be eventually harmed.

The damage caused by free radicals is cumulative.

Cell membranes serve a very important role in cell protection, transfer of information and in presenting surface molecules that, like *IDs* announce the body's guards which type of cell they are and what their function is. When the lipids that make up the cellular membrane are oxidized by free radicals, communication among cells is interfered with and this manifests in a deficiency in the functionality, including the destruction of protective fats in the cell membrane and liquid retention. This, in turn, accelerates the aging process.

These harmful effects from free radicals are being researched as causal factors in diseases that affect the nervous system, such as Alzheimer's. Also being researched are the effects on the immune system. The body reacts to free radicals by trying to repair the damage, but even its repair and regeneration capacity could be compromised by such substances.

Degenerative and proliferative processes, including aging, are explained by the above-mentioned phenomenon known as cellular oxidation. If free radicals attack the molecules that participate in cellular reproduction, cells can become cancerous. They can also harm those cells responsible for removing the cholesterol from the blood which would allow the formation of plaques within arteries and cause coronary disease.

The same way that the human body is multidimensional, so is the Universe in which we live. Therefore, there are different kinds of stressors and their impact could manifest in one or more dimensions of our existence, as can be observed in table 1.

Table 1. Types of Stressors

Stressors	Description
Physical, mechanical, biological	Traumas, starvation, sunstroke. Lack of sleep, overexertion, too much darkness, too much light. Extended periods under artificial light, work overload. Parasites, viral or bacterial infections.
Chemical	Pollution of the water, the air, the food, the soil and the food with chemicals products. Medication.
Electromagnetic	Radio waves from radio receptors, TVs and cell phones; Low frequency electromagnetic fields, computers, electro-domestic devices, planes.
Emotional	Situations that elicit a stress response from the body (conflictive relationships, job and study challenges, health conditions, financial hardship, losses) News in the media, traffic, recession, war, immigration status. Emotional abuse.
Mental	The above situations continue to stress us when we harbor disturbing thoughts about what has happened or may happen. Preoccupation and conflicting thoughts translate into anxiety, fear, anger.
Spiritual	Living what feels as a meaningless life. Quest for balance, meaning and purpose can be stressful for some people at certain times.

Chemical stress: pollution of water, air, and food

According to the United States Environmental Protection Agency (EPA), total U.S. emissions rose by 16.3 percent between 1990 and 2005. "As the largest source of U.S. greenhouse gas emissions, carbon dioxide (CO_2) from fossil fuel combustion has accounted for approximately 77 percent of global warming potential (GWP) weighted emissions since

1990, growing slowly from 76 percent of total GWP-weighted emissions in 1990 to 79 percent in 2005," the EPA stated in the *2007 Inventory of U.S. Greenhouse Emissions.*

From 1990 to 2005, says the same report, transportation emissions rose by 35 percent due in part "to the stagnation of fuel efficiency across the U.S. vehicle fleet."

As the chart bellows shows, the United States, with only less than 5 percent of the world's population, contributes close to 25 percent of the total pollution with carbon dioxide. China and the European countries are the other two main polluters of the world.

	The USA	EU Countries	China	Total
Population of world:	4.6%	6.3%	21%	31.9%
World economy:	30%	23%	3.2%	56.2%
CO_2 emissions:	24%	14%	13%	51%

Source: www.vexen.co.uk/USA/pollution.html#Pollution. Retrieved on Feb 6, 2008

On the other hand, industrial and domestic waste that ends up in water sources compromises the quality of the water polluting fish that are used for human consumption. Chemical additives that are used in food to improve flavor or preserve texture and maintain appearance for months, end up in our cells and their residues in the water we drink. More than 100 different pharmaceuticals have been detected in lakes, rivers, reservoirs and streams throughout the world. High levels of heavy metals in the fish, such as lead and especially mercury, have been reported and have generated sanitary alarms in places like the Gulf of Mexico and the Everglades.

There are some places on Earth where it is recommended that children and pregnant women should not be fed seafood. Mercury has been associated with the increase in the number of autism cases[13].

The *Food and Drug Administration* (FDA) and the *Environmental Protection Agency* advise women, especially mothers to be and the ones

---

[13] The medical journal *Health & Place* reports that children living next to a mercury polluter have a higher risk of developing autism. (See www.safeminds.org, May 1, 2008. Retrieved on May 18, 2008)

who are breastfeeding, and small children to avoid eating some types of fish and to choose fish and seafood that are low in mercury levels. These agencies also recommend avoiding shark, swordfish, tuna and tilapia. Information and warnings on the most contaminated places – that tend to change with season – are found in www.epa.gov.

When we pick a fruit at the supermarket, we unconsciously do so based on its appearance. Nature figured out how to let us know, through a color or smell, when a fruit is ripe and therefore it helps us get enough minerals and vitamins in a diet made of foods of different colors. Nature, though, is not limited to attract us visually toward those foods that are convenient. It also attracts us with flavors: a nutritional food is usually delightful to the palate.

A few years ago, they discovered how to turn coal-tar into brightly colored liquids that we now find in almost everything we use, from shoe polish to food. Researchers are very close to imitating natural flavors to perfection. For food producers, this is good news. Colors and flavors sell their products and they know it.

The problem is that a great number of colorants are derived from petroleum and flavors could be based on hundreds of natural or artificial chemicals. One would say, "And what is wrong with a little shoe-polish in a candy? What does it matter if a cherry Popsicle does not contain any cherries and the grape juice is not even close to a real grape? Do we really need to know what is in the cereal we serve to Tommie every morning?" Yes, we do.

Cigarette manufacturers have figured out how to use substances such as ammonium to improve taste and reinforce addiction. The same manufacturers have discovered that sugar and substances, like monosodium glutamate (MSG) added to food also promote consumption (addiction). This explains why we tend to finish one box of (certain brand's) cookies in a few minutes. Addiction prevention starts during childhood when children are taught to respect and love their bodies by eating healthy and controlling their compulsion for some foods. We should not reward them with junk!

It is true that the body has a great ability to tolerate exposure to dangerous substances, but the effect of these substances is accumulative and increasingly weakens our organism. The obesity epidemic in United States, for example, is attributed in part to MSG secondary effects and also to high fructose corn syrup – which is cheaper and today is used to sweeten most drinks.

We currently live in a world where neither the water nor the air are pure; foods are contaminated with pesticides, antibiotics and growth hormones, and have been processed or refined so much that they have lost the essential nutrients and fiber. If we also mention semi-synthetic food that has been dyed artificially and made with artificial flavors and preservatives, we should not be surprised that our body reacts by altering its functioning.

It may be true that Tommie is more sensitive than Joey and presents a stronger reaction to this chemical combination. For example, stomachache, bed wetting, muscle weakness, earache. He may also have an emotional reaction, low tolerance levels, hyperactivity, aggressive behavior and loquacity. Or he may have problems paying attention at school, reading, remembering words, solving a math exercise or writing. Generally, it is difficult to establish a direct relationship between symptoms and additives, and thus there are no specific laws to regulate the use of the additives in food.

A high exposure to chemical products or a deficit in compensatory factors has made many *normal* children sensitive to some synthetic substances found in the food or the environment. For the *Joeys* of this story, consequences may be less severe. To start solving *Tommie's* problems it is necessary to eliminate chemicals from the diet. It is necessary to learn to read the labels in the packaging of the food and, just in case, choose a diet based on fresh, organic and natural foods[14].

Electromagnetic stress: Low frequencies

We are constantly exposed to electromagnetic fields that are invisible. Technology has introduced electrical appliances that produce *Extreme Low Frequency* (ELF) vibrations that interfere with our own body's electromagnetic field.

The waves that transport information through TV antennas, computers and cell phones, appliances and airplanes are all sources of stress. Exposure to radio waves can cause serious damage to our organism. Our cells are so sensitive that the metabolism, the activity of those substances that transmit information, the synthesis of genetic material and the cellular behavior in general, can be affected by

---

[14] Organic is a term used for food that has been certified as being free of chemicals products. Natural refers to food that has not been industrially processed, or as not added colorants and preservatives.

unnatural frequencies, which increase the presence of free radicals in the body. We already saw what this can lead to.

This type of low frequency radiation, which is generated among other things by the cute radio-clock we keep next to our beds, also interferes with the proper communication between the nervous system and the muscles.

Our body tolerates really well the normal radiation in a house: electrical cables and electrical appliances. When adding cell phones, computers, microwaves, high tension electrical cables, and subterranean waters[15], it becomes too much and the negative effects from stress start to show up.

Recommendations to avoid electromagnetic fields:

> √ *Choose to live at least 50 yards away from high power lines.*
> √ *Sleep away from overhead wires and radio beams.*
> √ *Use a wind-up watch.*
> √ *Don't use electric blankets.*
> √ *Have your TV or computer away from your bed. At least, unplug them while not in use.*
> √ *Sit at least nine feet away from the TV.*
> √ *Avoid equipment that relies on radio waves or emits radiation of any kind.*
> √ *Use copper coils or semi-precious stones on top of fax, computer, TV, VHS and microwave.*
> √ *Build the body defenses through proper nutrition. Antioxidant, reduced fat diet is the ideal.*
> √ *If possible, have an LCD monitor or an anti-radiation screen on your computer.*

Emotional stressors – very real

Even emotional stress can contribute to cellular oxidation by stimulating the production of hormones that generate free radicals. When the liver tries to detoxify the body from these substances, it also generates radicals.

Almost everyone believes that the most predominant and harmful stressors are emotional; however, stress covers a wide range of

---

[15] Subterranean waters are rich in Radon, a gas associated with lung cancer.

situations and it attacks us from different fronts. News about national economy and international conflicts; societal hierarchies; high competition in our schools or jobs; our doubts about our job stability and even the traffic… all these factors stress us out every day.

There are other emotional stressors that add to the daily weight and make our body turn on the alerts, such as the loss of a significant one, divorce, moving or becoming a caregiver for a loved one.

Family and social roles have changed so much in recent decades that often, relationships become an important source of stress, instead of alleviating it, which should be their main purpose.

We all have had to face tough situations throughout our lives. The way they affect us depends on various factors. First of all, there is the intensity of each situation. Also, our perception of the situation and subsequent experiences reinforce or compensate the trauma's impact.

In the movie *K-Pax*, by director Ian Softley, the main character faces the loss of two of his loved ones, wife and daughter, and he responds by creating a delirious world within which it is possible for him to continue living.

In *The Prince of Tides*, by Barbra Streisand, each of the brothers implicated in the same drama, have completely different reactions to the trauma they all went through as children.

An email that circulated on the Internet contained a short story that illustrates what life might make of us. A chef's young daughter is overwhelmed with the first difficulties she experiences in life. Her father puts water to boil in three different pots as he is listening to her. In the first pot he puts an egg; in the second one, a carrot and in the third one, after the water boils, he mixes coffee.

After a few minutes, he shows his daughter what has happened in the three different pots and concludes: "What life does to us depends on us. It can turn us soft as boiling water does to the carrot; it can make us hard, as happened to the egg, or it can transform us into something new, with delicious aroma and flavor, as happened to coffee."

Exhausting thoughts

For the most part, we live *in* the mind. We anticipate the future, resent the past, and weave dreams that are not anchored in reality. Carl Sagan says in his book *Dragons of Eden*, that the price we pay for being able to anticipate the future is the disappointment it engenders.

I often think that one of the main philosophical problems is that we human beings have formulated the wrong fundamental questions.

Osho[16] tells an anecdote about Picasso. A gardener was watching Picasso painting on a beach, but he could not figure out the meaning of his painting. "I have been watching you paint," the gardener said. "You were so absorbed, so totally in it, that I was afraid to disturb you. Now that you have completed it, I cannot resist my temptation to ask, 'What is the meaning of this painting?'"

Picasso replies, "You ask me what the meaning of the painting is. Can I ask you - what is the meaning of the roses?"

Life just is. However, we insist on creating a meaning around it. We each find a different one. This is reflected in a popular Latino saying, *Each one talks about the fair depending on how they enjoyed it.* This happens because ours is a selective perception and that means that we do not perceive all there is (at least not in a conscious way) but rather just a fragment of reality. This selection depends on different factors (or filters if you may): how sharp our senses are, our vantage point (where we are located in relationship to what we are observing), our psychological or physical condition (mood, health, attitudes, values), our past experiences, our current needs and our goals.

We seem to choose to be exposed to experiences that reinforce our pre-existent ideas, confirm what we already know and validate our identity. How we justify what we do depends on our need to protect, maintain and validate both our self-image and self-concept.

Our senses receive a number of stimuli higher than the number the brain is capable of processing. For example, the eye can manage close to 5 million bits per second from which the brain consciously processes a very low quantity. That is why when in the process of assigning a meaning to our experiences, we automatically choose where to focus our attention.

To sustain this, I invite you to read the following:

THE interesting CAPACITY OF now tomorrow THE BRAIN TO you çan CHOOSE or not count JUST on the ONE depending AMONG on you to A SET OF feel in a given STIMULI IS AN because you never INTERESTING know exactly CHARACTERISTIC OF how turn out WHAT IS as a source of KNOWN because AS the news SELECTIVE partially what is PERCEPTION.

---

[16] World-renowned Hindu Master. He didn't personally write any books. Books by Osho are transcriptions of his speeches. See references at the end of this book.

What did you pay attention to? What elements did you find distracting? At the end of the paragraph, had you already chosen a series of words you wanted to read?

Our comprehension is selective as well. We need to discern the information that is not consistent and only retain that which seems good to the image we have of ourselves. That explains why, when experiencing something that we do not like, we avoid repeating it in the future.

Up until this moment, I have mentioned several stressors and factors that affect our way of reacting to stress. Those have become part of our daily lives because we have accommodated ourselves to them and accepted the status quo; lifestyles imposed by social, cultural and economical factors. We've been taught that that is the ideal way of living. It is even called a *dream life* that is sold to the whole planet on the name of progress.

We are submissively accepting the *globalization* imposed by the corporations that rule the world, in terms that have caused changes to life on Earth whose consequences we cannot yet foresee. If bio-diversity is essential for survival in nature, what will happen with Man when media, education and war finally accomplish their homogenizing hustle and we all end up adopting a lifestyle in which we produce more than we need, eat more than we really want and the body's responses are interfered with by numerous stressors from which we cannot just remove ourselves? Do we really want such a costly *global village*? Values, beliefs and behaviors that are essential survival mechanisms found by different cultures in their evolution, are dying. As the planet *progresses*, there is increasing devastation at all levels and this, on turn, affects the functioning of our bodies.

But the consumer of the products of devastation is almost as responsible as the one who produces them, the one who designs the course of action and the one who implements the strategies. There is, however, a power difference.

Testing balance

The body is constantly subjected to external and internal tensions (stimuli or stressors) to which it responds in a dynamic way, initiating an adaptation process. The goal is to keep certain stability and constancy for the body's inner conditions to, in turn, secure our survival. This process is in charge of what we've called the intelligent *inner healer* and it involves all the body systems.

If there are excessive tensions and the resources to respond to them are deficient, a crisis will develop. When this happens, the body seeks and holds onto all possible resources. For example, if a wound causes a hemorrhage, there will be a general vasoconstriction (narrowing of the blood vessels) that will then make blood flow slower in order to allow platelets the opportunity to gather together and form a blood clot to prevent a more serious blood loss.

During a fasting, sugars from our body's reserves are freed to continue responding to our needs. If fasting extends, the body responds by decreasing its energetic needs and sending any caloric intake to "fat storage." This explains, among other things, why we gain weight so easily after a strict diet.

At a psychological level, responses are similar. Our first reaction when they tell us a loved one has passed away is to resist believing it, going into denial. Often, almost simultaneously, we become irritable and angry. Denial and anger are mental mechanisms needed to process the loss, allowing for time to regain energy to face the tragedy.

In the Chinese language, the word *crisis* is represented by two characters; one means danger and the other one opportunity. This synthesizes the two poles of such moment: the danger to succumb in the face of a loss and the opportunity to evolve to a new state. Crises offer learning opportunities for the body in all its dimensions.

The same way that cells involved in defending the body during an infection *remember* the invader, our mind keeps what resources it gained during testing times. If the germ ever comes back, these cells can make the immune system respond quickly. The loss of a loved one strengthens us spiritually by showing us possibilities that we had not perceived before, modifying our roles and some of our behavioral patterns.

Communication systems in the body provide needed information on the changes that occur within the organism or environment. The body then can evaluate systems and resources while facing a threat by returning to a former state or evolving toward a new state in which it will be more efficient. The balance achieved is called *homeostasis*.

If existent resources are not sufficient to attend to the crisis or if the circumstances that caused the crisis prolong over time, there is a possibility that a dysfunctional adaptation can occur, which is also a way of keeping certain balance. This *maladaptation* manifests with symptoms and later with what we know as disease.

In both cases, there is a dynamic process, an *intelligent* response, which seeks above all to preserve function and save resources to guarantee survival. Critical conditions have to be extreme or the body has to exhaust its resources before it succumbs.

For example, symptoms like fever or diarrhea are both signs of alarm and responses of the body to an emergency. When suffering from an intestinal infection, fever causes the metabolism to speed up and white cells to increase their numbers in order to keep the invader at bay. Fever also increases the production of substances such as *interleukin* and *interferon*, which promote white cell activity.

When facing an alert, the body responds to protect itself from both an invader and the toxicity or deficit that results from it. Diarrhea – unusually frequent or excessively watery bowel movements of intestinal waste – constitutes a defense that facilitates and speeds up the removal of bacteria and toxic substances from the body.

In both cases, fever and diarrhea, there is an excessive loss of body fluids. By producing an intense thirst and dryness of the mucous membranes of the mouth, the body makes sure that we feel thirsty so that we substitute fluids.

A certain amount of tension (normal stress or *eustress*) necessary to stimulate our functioning, leads to inner *homeostasis* (physiological) and adequate management of the external environment (psychosocial). If we lived protected in a golden cage where all our needs were satisfied even before expressing them, we would not know how to adapt to the external world once released from the cage. We would not even be aware of our needs or how to express them.

Domestic animals or those that have always been captive in zoos undergo adaptation to the captive environment, rendering them less well adapted to survival in the wild. And, I think, this is now beginning to happen to human beings. We no longer understand what our bodies are trying to tell us. We visit the fridge when we need a hug. We eat when the body is requesting liquids.

One of my concerns is whether or not modern lifestyles put our capacity to adapt at risk; if they will, in the long run and permanently, compromise the innate body's intelligence; if they will make us completely dependent on external agents or on instructions to satisfy our needs. From the evolutionary perspective, our increased distance from nature has resulted in the loss of instinctive responses. In most cases, instinct has been replaced by knowledge either at the emotional, sensorial, mental or cultural level. A good example is that human mothers are the only ones in the animal kingdom that need instructions on how to breastfeed their offspring.

If we mostly inhabit artificial climates, look through tinted glasses that change natural colors, use too many antiseptics to kill normal bacteria and predetermine our behavior by learning that which is *right*, will the human body loose its ability to adapt adequately to cold or heat? Will our pupils loose their ability to contract when in the presence of excessive light? Will we produce enough antibodies to identify the external antigens when everything around us is disinfected and we have not been exposed to such antigens? Will we be in a condition to respond in a creative way to unexpected adaptation challenges?

If we entrust our memory to personal digital assistants, cell phones or any other electronic system, will we end up not needing it until we loose our ability to memorize at all?

Even though Darwinism argues that survival of a species depends on natural selection and this one is the product of chance, scientists have found some evidence indicating that adaptation changes are transmitted to descendants. Are we threatening survival of the human species with our lifestyle?

Optimists trust that the homeostatic capacity of the body is infinite. Pessimists predict our extinction. Perhaps the truth lies somewhere in the middle, maybe in the knowledge that humanity has accumulated over the years to find solutions in more harmonic lifestyles that are more respectful of the body and nature.

Life as a luminous halo

Life is not a series of gig-lamps symmetrically arranged; life is a luminous halo, a semi-transparent envelope surrounding us from the beginning of consciousness to the end, Virginia Wolf said, to explain the peculiar features of her writing.

As a character in one of my previous writings, and based on what research told me about her, she also said:
"I believe life is far from being this way (mechanic). The mind receives thousands of impressions. Here, while we talk, you perceive my dress, the door that just opened, the waitress circulating across the room and, also, you listen to me and to

the background music; some impressions will be so subtle that they will poof away and some others will remain recorded with the strength of steel. It is like an atom rain: Life is not a series of gig-lamps symmetrically arranged; life is a luminous halo. "[17]

In order to conceive of the body in its multidimensionality and in its vibrational expressions, to understand how the inner healer works and to recognize the communication that takes place at different levels of our being, we need a vision that is holistic, ecological and dynamic. We need to untie ourselves, even if only for a moment, from the frame of reference that humanity has built for centuries.

How do we change the lineal vision and thought process we are characterized for? Isn't this the type of thought process stimulated at schools where priority is given to logic, math and writing? How many Western schools consider, as part of their curriculums, things such as meditation to help develop intuition and perception? And how many schools train our senses?

Most of our languages are lineal – Chinese and Japanese are not. Language is composed of a chain of minimum elements that combine and articulate in a hierarchal way. The type of thought process that predominates in the West is lineal as well, but life is not; life is paradoxical. Intuition is paradoxical, multidimensional – not lineal. It is the path and the result of our connection to that wholeness that we belong to.

Religions, philosophy and science have registered different views of the world throughout history: some are mechanistic (causality concepts), some dialectic, (dynamic concepts).

For example, basic assumptions in Galileo and Newton, predominant in the scientific world in the West in the last few centuries, come from the observation and manipulation of inert objects – not living, complex ones. They wanted to find out how a given cause would produce a given effect. They intended to isolate variables and reduce nature to its fundamental units in order to being able to study it. The intrinsic risk in their explanations of the world is that, without a context, human beings are reduced to their parts – Man is fragmented and, in the long run, sight of the whole is lost.

---

[17] Silvia Casabianca. *In search of my own room.* Essay published in: *El Pequeño Periódico.* No. 19. 1985. p. 8-9.

Quantum theory, formulated at the beginning of the 20th century, transformed physics and is helping us to understand (or remember) that there is a basic unity in the Universe (a word that means *unique and diverse*) as Hindus, Buddhists and Taoists have understood for centuries. Things that we perceive as individual are manifestations of that whole.

When it talks about the principles of polarity, the Kybalion postulates that unity exists within a whole that flows and changes:

Everything is dual; everything has poles; everything has its pair of opposites; like and unlike are the same; opposites are identical in nature, but different in degree; extremes meet; all truths are but half-truths; all paradoxes may be reconciled.

And according to the principles of rhythm:

Everything flows out and in; everything has its tides; all things rise and fall; the pendulum-swing manifests in everything; the measure of the swing to the right, is the measure of the swing to the left; rhythm compensates.

The Chinese express similar principles when explaining the Tao as a whole and polarity as Ying and Yang. *T'ai c'hi t'u* (or the symbol of the *ultimate supreme)* has been symbolized as a circle split into two opposites. The dark swirl in the symbol representing Yin – is counterbalanced by the white swirl representing Yang and each opposite is seen as containing the seed of the opposite. Yin is growing to become black, while black grows towards becoming white. The symbol is a metaphor for the active and passive forces in the universe.

In the Western countries, the development of a systemic epistemology was essential for the development of science. Since the fifties, this development has impelled medicine towards new approaches such as family medicine and family therapy, characterized by dynamic views.

General System Theory (GST), formulated by Ludwig von Bertalanffy, developed as a science of the living systems and focused on the characteristics of global systems. It amplifies the Aristotelian concept that states that *the whole is greater than the sum of its parts*. This theory produces a paradigm shift where the interest in lineal cause-and-effect relationships is replaced by the study of the whole, patterns and *feedback loops*.

GST says that systems – a series of structures related to each other in order to fulfill a function – keep conditions relatively stable thanks to the existence of feedback mechanisms that allow for self-regulation.

Feedback loops are defined as information processing mechanisms through which the system determines the origin of its current state, the origin of an environmental disturbance – or both – and generates a response. The response could have the purpose of taking the system towards the correction of any deviation or of the evolution to new a state.

The concept of feedback circuits is close to that of homeostatic control mechanisms, the processes in charge of keeping the system conditions constant while facing changes in the environment. Sometimes the function of these body circuits is interpreted from a mechanical perspective, equaling them to thermostats, although this explanation falls short. Inside a system there is always a free game of forces, a complex interaction.

For example, when facing extreme environmental temperatures, constancy in the body depends on factors such as our diet, serenity, amount of liquids we drink, skin vasoconstriction or vasodilatation processes, breathing rhythms, energy production at the cellular level and muscle work, among others. There is not a switch that goes on and off every time the temperature is not within the desired range.

For practicality's sake, we will hereby call *feedback loops* those dynamic interaction mechanisms that characterize systems and help maintain *homeostasis*.

The ovulation cycle is a great example of how feedback circuits operate. It is based on communication between pituitary, ovaries and uterus, whose changes are determined reciprocally.

The pituitary produces stimulating hormones that, every month, send a production order to the ovaries. These respond by elaborating substances (estrogen) that influence the uterus, which in turn creates conditions for the eventual nesting of a fertilized ovule. Once the ovary has produced enough estrogen, the pituitary stops its own hormone production. The uterus informs the pituitary of whether the nesting occurs or not, and this one responds by giving instructions to the ovaries about new productive goals. The ovaries then free the corresponding products (progesterone), which determine the desirable conditions that the uterus should now create.

Understanding and respecting these dynamic interactions is essential. It is the best possible argument to defend the need to be cautious about being exposed to disturbing elements such as medications, recreational drugs, chemicals added to food or extremely low frequencies – all able to interfere with the fine synchronicity required by the *inner healer's* intelligence.

Many women find their ovary function compromised after the long-term use of contraceptives. People whose thyroid hormone production is lacking are prescribed hormonal substitutes that do not change the gland's condition but rather generate dependency on the synthetic hormone, sometimes even for life. Frequently, the administration of a hormone not only interferes with the delicate network of reciprocal regulation between glands, but also masks deficiencies that, if corrected, would likely solve the problem. For example, an insufficient amount of iron and manganese could explain a decrease in the production of thyroid hormone or the normalization in the production of serotonin in a depressed person could result from increased physical activity.

When hormones come from external sources, normal cycles, regulated by *feedback loops*, are cancelled out. When administering synthetic estrogen, for example, the pituitary no longer feels the need to stimulate its production in the ovaries. The same could be said in the case of the thyroid. This is without even mentioning the undesired secondary effects that most synthetic products bring about.

From this perspective, products that are presumed to stimulate the normal functioning of the organism[18] have become popular. It seems that technology is making efforts to compensate the same problems that it has caused. Years will pass by before we know the long-term effects of the new products that are commercialized nowadays as safe and are being sold without regulations.

I hope we do not repeat the same mistake made when taking megavitamins became popular and nobody knew that overdoses could be harmful.

In November, 2005, the article *Nutrition from the Kitchen, not the Lab*, published by the *Health & Nutrition Letter of Tufts University*, confirmed something Chinese doctors have been warning about for years. Isolating essential nutrients of a plant to turn them into a pill does not guarantee that the pill will have the same effect as the plant.

---

[18] Natural products, supplements, herbs and vitamins that are sold with no need for approval by the FDA or the EMEA (regulate food and drugs in USA and Europe, respectively) are marketed with a warning stating that the claimed benefits are not supported by the sanitarian authorities. Both agencies are now considering to control the commercialization of those products, due to the fact that some of them (Efedra, for instance) have been found to be harmful.

Supplements and artificial vitamins do not replace a good nutrition or guarantee our health. According to studies mentioned in the article, counterproductive effects have been found. Vitamin E, for example, is needed in our diet but produces undesirable results for our health (nausea, gas, palpitation, tendency to hemorrhage) when taken in large doses as a pill.

Communication is vital for life processes

Traditional Chinese Medicine (TCM) and Ayurveda medicine are the two major legacies for health and healing from the ancient world. Both systems, which in the West are widely acknowledged as alternative medicines, talk about communication between energy centers, between the subtle and the dense parts of the body, between the body and the mind and between the body and the universe (heaven and earth for Chinese, universal energy and the body in Ayurveda).

In TCM, one of the functions of the meridian system is to promote communication between the internal organs and the exterior of the body, and to connect the individual to the rhythms of the earth and heaven. Other functions are to balance yin and yang energies (feminine and masculine complementary opposites) and to distribute Qi (also known as Chi, vital energy) throughout the body.

In Ayurveda, the chakras serve the purpose of receiving energy from the universe and distributing it within the body through "pathways" called nadis, connecting the physical, emotional, mental and spiritual bodies, to maintain balance.

These concepts seemed absurd to Western medicine (still do). Notwithstanding, in the past few decades, the idea that the organs communicate and that this communication is essential for wellbeing, has been studied, broadened, fathomed and confirmed.

Indeed, the organs communicate and this communication happens, from what we can measure at this point, at least in two ways: biochemically and electrically.

Certain molecules can be found in the surface of the cells that confirmed an old suspicion. Researchers had hypothesized that pharmaceuticals worked in the body because they found some kind of matching receptor. This is not a phenomenon exclusive to pharmaceuticals; on the contrary, it explains the existing relationships between the different organs in the body.

Candace Pert, from the *Department of Physiology and Biochemistry in Georgetown University*, D.C., compared these receptor molecules with scanners looking for compatible chemical substances (messengers) in the fluids in which they are submerged. Once receptor and messengers bond, the messenger can deliver its message.

The message has a regulatory effect over the cell. Some of the messages kick off the production of protein; others make the membrane more or less permeable. Some messengers support each other (synergism) while others deliver opposite messages (antagonism).

By proving the existence of opiate receptors at the beginning of her career as a researcher, Pert[19] contributed to unite endocrinology, neurology and immunology in one science (known as *psychoneuroimmunology*), thus ratifying the existing unity of body and mind. Her studies about the biochemistry of the body do not point to a power exerted by the mind over the body, or that thoughts dominate the body, but that the mind becomes the body. This might sound like a subtle difference, but it is not. When we think that the mind rules the body, we are emphasizing duality, not unity. Pert's work with what she called the *molecules of emotion* demonstrates the flow of messages throughout the body, messages that are transported by molecules. She calls *mind* the product of that process of information transmission.

Mainstream medicine has recognized that immune, endocrine and nervous systems own a common chemical language. Such systems are united by a multidimensional network, an ensemble of peptides that transport data. But the talks are not restricted to these three systems. Actually all bodily systems make up a dynamic and interactive net. All of them are agents under the command of the *inner healer* as they are involved in tissue regeneration, regulation and repair and participate to a lesser or greater degree in the communication process.

All of us at some point have experienced a state of exhaustion after suffering a bone fracture, a sprain or even diarrhea. Why is it that the whole body seems compromised when we have suffered a local injury? It is because the whole body mobilizes all of its resources to repair tissues that have been damaged.

We could say without falling into exaggeration that the liver participated in the healing of my injured knee. It probably provided some stored sugars that were needed to supply the increased demand

---

[19] Pert's peers received a nomination to the Nobel Prize in 1995, for the discovery of the opiate receptors.

for energy, necessary for the metabolism of the injured cells or the removal of dead tissue. The liver thus must also have sacrificed a bit of its blood quota during the redistribution of fluids that is normal during an inflammatory progression.

As mentioned above, our *inner healer* is subject to the efficacy of the communication network. What are then the elements that make up the communication network?

From an evolutionary perspective, communication appears the instant life is created. Recently it was discovered with great cheer that bacteria not only communicate among themselves, they also cooperate. This is how Bruce Bower gives account of the discovery in *One-Celled Socialites. Bacteria mix and mingle with microscopic fervor,* an article published in *Science News Online*[20].

> *They congregate in immense numbers to fend off enemies and the brute forces of nature, to obtain food, to reproduce, and to move to greener pastures. They're adept at forming bands to hunt prey, which are consumed on the spot. Vital messages repeatedly course through these assembled throngs. Under some circumstances, certain community members sacrifice their lives for the good of the rest. At other times, entire congregations cozy up to unsuspecting hosts before coalescing into stone-cold killers.*

Each and every living being, and also the whole social structure, is dependent upon its communication systems. As organisms ascend in the evolutionary chain, the communication becomes a more complex thing but the elements that grant it continue to be essentially the same.

All communication, inside and outside the human body, and in social life, imply interactive emission and reception of messages between two interlocutors. It presupposes a message but also the capacity to receive and correctly interpret the message. Furthermore, communication needs a media to deliver the message and when direct communication is impossible, a messenger is necessary.

Balance could break off when one or more of these elements fail, as can be illustrated by a game I used to play as a child. In *broken telephone* (also called *Chinese whispers*) each successive participant secretly whispers to the next a word or sentence whispered to them

---

20 www.people.fas.harvard.edu/~kfoster/BowerScienceNew2005.htm, Retrieved on February 11, 2008.

by the preceding person. When the message had lived the full circle, errors due to mishearing had accumulated, resulting in a message that greatly differed from the original.

In the same way, the human body is conditioned to the optimal functioning of the communication network. Sense organs, the nervous receptors on the surface of the body, chakras, communication *routes* (lymph, blood, synapses, nadis, meridians) are all elements that need to excel in performance to keep intact the transmission of information. Even when it goes unnoticed, any alteration of the communication system will have an impact on the totality of its function.

Although health sciences advance rapidly, sometimes they contribute only a fragmented view of the human body. When pain killers or anti-inflammatory medications are provided to a patient to alleviate him, the doctor is not working from the perspective of the whole. The pain killer or the anti-inflammatory may flood or saturate the receptors that might be essential in the process of granting the resolution of the inflammatory process.

In my experience practicing reiki and Trager, I have seen that conditions tend to become chronic after being medicated, while in most cases the resolution of an injury or a condition is resolved faster with proper nutrition, relaxation and neuromuscular reeducation[21].

I have thus come to the conclusion that a therapist, whatever the dimension she is working with, should be totally respectful and encourage respect for the communication processes taking place within the body. Even a massage, which most people deem innocuous, is a stressor that requires a response from the body.

Any external intervention to the body risks balance. That's why we need to move towards preventive care and self care, which require health providers to inform and educate clients.

By getting to know our bodies really well, we will also learn to take responsibility for ourselves, become experts in our own body and prevent symptoms or flares.

---

[21]Anti-inflammatory medications rofecoxib (Vioxx) and valdecoxib (Bextra), for instance, were withdrawn from the market in 2005 when it was discovered that they increased the risk for cardiovascular disease. The cleaning of the arteries supplying blood to the heart also depends on the immune system, whose function is delayed when anti-inflammatory medication is administered.

Everything vibrates

Throughout the book we have insisted and will continue to insist in the central idea of a multidimensional body. We exist in a simultaneous way in different spheres: biological, psychological, social, spiritual and cosmic. Even though the physical is a tiny part of who we are, the physical draws all of our attention because it's visible, palpable, tangible, dense. However, the density of the body is a product of the way our atoms and molecules are packed up and of the speed at which their electrons spin.

To understand this, imagine the blades of a fan, which are visible and discernible only when the fan is not turned on. As it starts spinning, the blades seem to create a homogeneous disk and you can no longer distinguish one blade from the next. The smaller the fan, the stronger the sensation that there is no space between the blades.

We visualize ourselves as dense bodies because of the speed at which electrons spin. Actually the distance between the atom's nuclei and the periphery cloud of spinning electrons is proportionally larger than the distance between The Earth and the sun, pretty much as we see a solid disc and not single blades in a spinning fan. In the space between the nuclei and the spinning electrons there exists an electromagnetic field that probably stores information.

We know that the *subtle* dimensions of the body exist, even if they are invisible to most people. Who doubts that we have a mind, even if nobody has been able to find where it resides. One challenge found by psychotherapists in their practice is that it is actually impossible to do objective measurements of mental and emotional phenomena, at least not in the way medicine tests a drug. I see us, psychotherapists, as people who deal with subtle aspects of the body.

I think that the main characteristic of multidimensionality is that the same phenomena have simultaneous but peculiar expressions at each level. A person who was betrayed by a friend will have unpleasant thoughts about the incident. When he concentrates on these thoughts he prompts an emotional reaction (rejection, rage) that on the physical level will be expressed in the production of stress hormones. At the energy level, it might cause a blockage and at the spiritual level, his emotions and thoughts might compromise his capacity to trust other human beings.

The notion of subtle bodies comes from different philosophies, esoteric practices and oriental medicines. However, many Western schools of thought and many individuals have long ago adopted this concept. Some consider subtle bodies to be as layers one on top of the other: actually a lineal concept, which ends up creating a separation between one layer and the next.

In a holistic perspective, it seems more appropriate to use the image proposed by Osho, who in talking about subtle bodies considers that they interpenetrate each other in a similar way that a sponge takes in water; the water flows in through the pores and the air flows through the water and the sponge.

Another good example is to imagine a container that we fill with tennis balls. To fill in the free spaces between the balls, we add golf balls and then marbles and finally sand and water until the recipient looks really full. There is still space for the air that circulates through the water and surrounds the container connecting it with all that is. Each element represents a dimension of the multidimensional body.

The sponge, the water and the air, as well as the different objects in the container, would have different densities and dimensions, the same as the different dimensions of the body have each a different vibratory frequency.

All is vibration. In us, as in the universe, all vibrates and, depending on the frequency and length of the waves, the manifestation to our senses will be different (color or sound). There are also vibrations that science has been able to detect (ultraviolet rays, ultrasound) even though they are imperceptible to our senses.

Electromagnetic spectrum is the name that scientists assign to the group of various known radiations. Radiation is energy that travels through waves and spreads like the light of a bulb or radio waves. Color, sounds, microwaves, infrared and ultraviolet light, x-rays and gamma rays are all vibrations. The difference between them is in the number of cycles in a certain time unit (frequency) and the length or amount of energy that each contains.

Electromagnetic radiation can be described as a stream of *photons* – particles that have zero mass, that create waves when moving at the speed of light. Each photon contains a certain amount of energy and all electromagnetic radiations consist of these photons. The only difference between the diverse types of electromagnetic radiation is the amount of energy found in the photons. Radio waves have low energy photons while the ones found in gamma rays contain the most energy.

We also emit different types of energy that is perceived as heat, electricity, light, sound or magnetism.

I suggest a simple exercise:

Raise your hands above the head and pat yourself from shoulders to hands. Now go from your hands to shoulders. Repeat. With your hands still up, open and close them for a few minutes. Shake your hands. Now lower your arms and with your fingers together, forming a concave shape, put one hand in front of the other one, close your eyes and feel. After a few minutes, separate your hands slowly and try perceiving which changes take place. Put your hands together again. With your hands apart about 5 inches look at the floor through the space between the hands. Now separate your hands and lower them, and look at the same spot on the floor. Is it different?

Quantum physics found that atoms are formed by *quarks*, particles that are even smaller than electrons and protons. It proved that, at the photon level, there is a duality wave-particle. The subtle and the dense, energy and matter, are opposites of a unit of opposites, or better said, equivalences of an equation – the infamous equation formulated by Albert Einstein 100 years ago: Energy equals matter times speed of light squared ($E=mc^2$). Opposites are not exclusive of each other, they form a unit.

Electrocardiograms, electroencephalograms and electromyograms are all medical procedures that register electrical activity (waves or vibrations) of the heart, brain and muscles, respectively, showing that our bodies generate electricity. An expert may know, for example, what is wrong with the heart when he registers the electrical or magnetic field produced by this organ.

Before quantum mechanics revolutionized physics, three important researchers, Michael Faraday, Nikola Tesla and Thomas Alva Edison, came across during their experiments with the phenomenon of electromagnetic fields surrounding the human body[22].

---

[22] Faraday discovered that electrical current could be induced by the use of magnets, which opened the doors to energy generation (which explains why our homes today are invaded with appliances). I we can thank Tesla for wireless communications. Edison created, among other things, the phonogram and the electrical bulb.

Several studies have reported that healers produce a magnetic emission in the palm of their hands and its effects (for example, on how plants grow) can be observed even if the detected field is weak.

The SQUID (Superconducting Quantum Interference Device) is able to measure the biomagnetic field generated by a single heart beat, a muscular movement or the activity pattern of a brain cell. It is being used in research in several medical centers around the world. The biomagnetism registered by the SQUID is similar to the one healers have perceived for centuries.

The brain produces signals of about 10^{-9} gauss[23]; the heart of about 10^{-6} gauss and while they are healing, the healer's hands can produce signals of about 10^{-3}, 1,000 times stronger than the heart's[24]. The magnetic waves surrounding the body can also be measured with the SQUID.

Kirlian photography or electro-photography, is taken in the presence of high-frequency, high voltage and low amperage electrical fields. It was invented in 1940 by Russian researcher Seymon Kirlian who proved with images the existence of a field or aura around bodies and objects, a changing and dynamic field that is affected by external and internal circumstances.

In more recent times, it has also been established that cells absorb light and then emit photons from the cell nucleus, from the DNA[25].

This light is too weak to be seen by the human eye but it has been detected using laboratory techniques. These particles have been called *biophotons* and constitute a low-intensity radiation that all living systems emit in the spectrums of visible and ultraviolet light, and is correlated with most, if not all, physiological functions. Because body phenomena are not unidirectional, it is probable that *biophotons* circulate from the DNA to the periphery and vice versa, interconnecting the body. *Biophotons* have been related to the chemical reactivity of the cells, growth control and biological rhythms. This would explain, at the subtle level, the communication between cells, tissues and organs – a communication that has been proved to exist by biochemistry and about which we'll talk about more.

---

[23] Gauss if the unit of measure of magnetism.

[24] Liu, Y et.al. *The effects of Taoist Qigong on the Photon Emission from the Body Surface and Cells, Proceedings of the First World Conference for Academic Exchange of Medical Qigong*, Beijing, China, 1998. In www.chiexplorer.com/newsletters.

[25] DNA: Deoxyribonucleic acid, where we store genetic material that determines our phenotype.

Among scientific explanations on the existence of luminous radiation is the one that sees the *biophoton* emission as a byproduct of biochemical reactions at the cellular level. Another theory considers that these photons constitute an unusual form of light, coherent light, similar to laser, which would give credibility to the idea that there is an intelligence behind biological phenomena (an aspect of consciousness). Applications are being studied in the food industry, medicine, pharmacology and environmental sciences. *The Technical Universitat Ilmenau* and the *International Institute in Biophysics* in Germany are pioneers in research in this field.

A few years ago, the Chinese government sponsored research with healers that practiced Qi Gong. To the surprise of many, they found that the hands of these healers emitted a special vibration that corresponded to sounds below the audible spectrum. Devices that have been successfully marketed in the United States are based on the findings of these studies according to which the subsonic waves increase circulation, reduce swelling, alleviate pain and promote scarring. Liu Guo-long M.D. and Richard Lee from the *Beijing College of Traditional Medicine* confirmed these findings recently[26].

In the market, you can find different devices that combine light and sound in order to induce relaxation, meditative states and that also control stress (brainwave entrainment).

Physics tell us that all materials known to us could have magnetic properties. If they don't, it is due to the fact that the magnetic fields of the different atoms are not *aligned* and their charges cancel each other out.

On the other hand, when these forces are aligned in the same direction, materials show magnetic properties. It is like the *tug-o-war* game: tension is kept and the rope stays tensioned while the teams pull the opposite way. Once a team either pulls harder or gets weaker, team members fall on top of each other, *attracting* the opposite team to their side due to the hard pulling.

In the case of materials such as iron, its magnetic fields are aligned in the presence of magnets. We are still far from having the necessary mechanisms or devices for proving if a healer, in some cases, can achieve a certain vibrational frequency, *aligning her atoms* (or her

---

[26] *Infrasonic Simulation of Emitted Qi from Qi-Gong Masters,* Published online February 2006. www.drwastl.org/files/eeg_and_qgm.htm

consciousness or her energetic fields) in such a way that she can generate a field that would influence, like the magnet does to iron, the *aligning* of the atoms of the other person. But the hypothesis is an interesting one.

Researchers have found that the brain waves of a healer change whenever she is treating a person, inducing similar brainwave frequencies in the receiver. Moreover, the frequency of these waves is similar to those that are detected on Earth, making us wonder if healing is then a resonance phenomenon. (Superposition of waves, constructive interference).

Resonance is the quality that a body has to rebound a vibration, forming a wave that becomes a circuit. The circuit created between healer and receptor would allow for a maximum flow of energy.

People who receive reiki frequently report a heating sensation that is more or less intense in different parts of the body. Healers report a similar sensation of magnetic attraction between their hands and the other person's body. This sensation perceived in the hands during the healing act is explained in reiki as the product of the encounter between Rei and Ki.

Michael Shea[27] compares the resonance phenomenon with that known as induction in which a conductor (in this case the person receiving) is electrified when near to a body charged with electricity (the therapist).

James and Nohora Oschman, from Nature's Own Research Association in Dover, New Hampshire, say the following in their article *How Healing Energy Works*[28]:

> *Seto and colleagues have found that practitioners of traditional health and martial arts exercises, including Qi Gong, Yoga, meditation, Zen, etc., are able to emit very strong pulsating magnetic fields from the palms of their hands. The fields are so strong that they can be detected with a simple magnetometer consisting of two 80,000 turn coils of wire connected to a sensitive amplifier. The fields are about 1,000 times stronger than normal human biomagnetic fields such as the magnetocardiogram studied with the SQUID.*

---

[27] Body worker who has been an instructor and author of *Somatics*.
[28] This article was published for the first time in the summer of 1993 in *Convergency Magazine*. Volume 6, 3. p. 24–30.

The frequency and strength of the pulsations recorded by Seto and colleagues are most remarkable. The pulses occur from 4 to 10 times per second. This is an important frequency for several reasons. First, it is in the same range as human brain waves as detected in the electroencephalogram. Secondly, it is similar in frequency to biomagnetic pulses recorded from the hands of a therapeutic touch practitioner by Dr. John Zimmerman, using a SQUID magnetometer. Thirdly, the pulsation frequency varies from moment to moment. The Earth's atmosphere also has variable electric and magnetic oscillations in the same frequency range.

The phenomenon of our body's density is even more interesting if we take into account that the distance between the nucleus of those atoms that form our matter, and the place in which the electrons circulate, is proportionally greater than the distance between the Earth and the sun. We could say that we are made of more space than of particles! And, what is there in that space?

Remember school science fairs where there always have an experiment to prove that, around an electric wire carrying current to a lit light bulb, there is a magnetic field that is visible as light in the darkness.

The relationship between magnetism and electricity was accidentally discovered about 100 years ago by Hans Christian Orsted who noted that a compass needle changed direction whenever it was near a wire conducting electricity.

From basic chemistry, we know that the protons in the atom have a positive electrical charge and the electrons have a negative electrical charge. Resting electrons are pure electromagnetic fields and there are electromagnetic fields around protons and electrons. Television and tape recorders are a good example of how sound and image can be transmitted and stored using electromagnetic patterns. Similar phenomena are found in the body.

Can that electromagnetic field around protons and electrons store recordings of images, sounds, smells and every other element that make up our memories? Researchers seem to agree that memory is not found in only one place. Specialists of the nervous system continue to be surprised by the finding that extensive injuries of the brain do not compromise the person's memory as it would be expected.

There are still no studies, regardless of how advanced they are, that let us know for sure where mind (or consciousness) is located – that is, thought, perception, emotions, will and imagination. We know a lot about the brain but hardly anything about the mind.

It is true that magnetic resonance equipment shows cerebral activity every time we experience an emotion, desire or when we have certain thoughts, but that does not assign the mind a specific location within the body.

Different neuroscientists define mind differently. To NASA neurologist Rodolfo Llinas, mind is co-dimensional with the brain, a product of its activity. In general terms, researchers agree that mind cannot be located. It is the byproduct of processes generated by perceptions that arrive in the brain as information and it comprises phenomena such as thought, memories and emotions.

It is interesting to note that, for Zen practitioners, mind is in the stomach. The brain seems to be the host of our intellect but, where is the one who thinks? They establish a difference between mind and intellect. Intellect's counterpart is intuition and they say this one seems to come from our guts. Intestines contract when we get emotional, according to Zen practitioners. We can squeeze our brains in search of answers but when we stop making the effort, when we give up, the searched answer often pops up in a dream or a poem we read randomly. It does not seem to come from an effort we made.

Osho talks about the cases of Madame Curie and Buddha – she was after the solution to a problem and he was looking for enlightenment. Both found what they wanted once they stopped trying. Madame Curie in a dream and Buddha while he slept under a tree's shade.

One hundred and twenty years ago, Professor William James and his student Walter Cannon, discussed these very same topics from a biological perspective. According to his studies, James defended that emotions came purely from the guts. Cannon found that there were specific areas in the nervous system that could correlate with emotions.

Twenty years ago, Candace Pert postulated that perhaps both of them were right because emotions produce changes in the body and the body acts upon emotions, in a two-way process. However, there is no specific place in our anatomy for our love for "Joe."

To sum up, there is enough evidence that proves that our bodies are vibration, energy: atomic, caloric, electrical, sonic, luminous and magnetic. The body is subtle and dense, matter and energy – multidimensional.

An energy network within a multidimensional body

Healers in different traditions and cultures were the first ones to detect the existence of energy fields and non-physical phenomena related to the body. Human beings have a special sensitivity that can be innate or learned.

There are interesting parallels between different schools of thought, all coinciding with the idea of an energy structure of the human body. For instance, the Hindu system of *chakras* and *nadis* is similar in many ways to the *meridian system* in Traditional Chinese Medicine and to the *Sephiroth* in the Hebrew cabala.

Buddhism, Hinduism, Ayurveda and Taoism have their root in ancient beliefs. All of them talk about a kind of energy (ki, chi, prana) that animates matter and circulates through vessels or conduits (nadis in the Hindu system and meridians and channels in the Chinese one). Hindus talk about *Sushuma, Ida* and *Pingala*, lunar and solar energy; Chinese talk about *governing* and *conception* vessels, *yin* and *yang*.

In most Japanese spiritual and martial disciplines, we hear about two types of ki (or seiki) converging in the *hara,* beneath the navel, and vital center of the body: *heavenly energy* comes through the crown in the head and *earthly or telluric energy* comes through the perineum. Some traditions add another source of energy, *ancestral energy*, emanating from the adrenals.

The main differences between the different traditions reside in their spiritual practices and the energy center they assign more importance to.

Hindus look for liberation from the cycle of death (Moksha), a *nirvana*, an existential state in which they can experience that all that exists, including us, is Brahma or absolute reality. The Hindu sage cultivates knowledge and contemplation, looking for detachment. He resigns the mundane reality and depends on others to make a living.

Buddhists renounce material attachments, which are considered the source of suffering. The prototype of the Chinese sage, on the contrary, is the Confucian gentile, a superior man who offers outstanding moral answers for family, politics or social issues.

Chinese Taoists, who look for the true nature of things, believe that we are born with a fixed amount of Jing (a certain kind of chi). They look for rejuvenation and longevity through alchemic practices that allow them to save chi or harmonize their individual energy with the universal energy. This is the purpose of martial arts. *Esoteric Taoism*, similar to the Tibetan tantra and to tantric Buddhism, looks to achieve through *internal alchemy* the transmutation of vital energy (sexual energy). There is again a similarity between the *Tan Tien* or alchemic cauldrons (inferior, middle and superior) in Tantric Taoism and sacral, heart and third eye chakras.

In Cabala, a symbolic map for the forces in the universe and their reflection or correspondence in the human body, known as the *tree of life*, consists of ten *Sephiroth* connected by 22 paths. Each *Sephira*[29] represents successive divine emanations that correspond to ten stages in the continuous evolution of the universe, Man and things manifested. Each Sephira is a seed that contains certain potential. The goal of a cabalist is to transform this potential energy in creation so that its free flow creates balance in our lives.

It is interesting to note the many similarities found between ancient and recent Eastern and Western philosophies and religions. Let me just mention the idea of life cycles and stage by stage development found in Sigmund Freud, Margaret Mahler and Eric Ericsson, to name only a few people who worked in the psychology field, which is an equivalent in the West to working with the subtle bodies (in this case, the psyche).

The main idea across these different beliefs is that our life is organized along a series of goals that we must accomplish successfully in order to develop our full potential. As we will see, the development of each chakra is also associated with an evolutionary task.

Eastern philosophers and Western esoteric healers support the idea of the existence of different dimensions in the body, assigning to them different forms and names. Many of them seem to agree that there are seven bodies, levels or dimensions; one of them is dense (the physical) and the others are subtle. Each one of them has a characteristic vibrational frequency that can manifest as color, from the lower frequency (physical) to the highest frequency (nirvanic or monadic).

Table 2 details some of the characteristics assigned to each body or dimension.

---

[29] Sephiroth is plural and sephirah singular.

Table 2. The different dimensions or bodies[30]

Body	Characteristics
Physical (dense)	What you can see, measure, weigh, touch
Etheric	Hologram of the physical body. Contains a blueprint that determines the form of the body. Described as having the same shape as the physical body but extending a few inches around it. Clairvoyants see it as a halo.
Astral/Emotional	Where desires and emotions dwell. Some clairvoyants have described it with the shape of an egg that changes from dark and opaque to bright and clear, changing colors according to feelings experienced.
Mental	The house of our thoughts. Osho divides it into concrete and mental and abstract mental bodies. The latter connects spiritual levels with former levels. Its size would change according to the kind of thoughts that it hosts.
Spiritual or Buddhic	Where love and unity are generated. It extends towards the universe.
Cosmic	Spiritual plane where we connect with the whole, the transcendent.
Nirvanic or monadic	Where permanence of being resides, the essence.

It is also said that there is a correlation between the development of the subtle bodies and the development of the chakras. According to Osho, at birth, the physical dimension is already developed and each of the other bodies or dimensions exist as a seed. It would take about seven years to develop each one of the dimensions with their own characteristics.

---

[30] The information presented in the table is taken from different sources: Hindu Taittiriya Upanishad doctrine; Neoplatonism, that considers the dimensions of the body, the spirit (neumo), the soul (psyche), and the divine intellect (nous); Vedic tradition; Cabala; la theosophical theory (Blavastky, Rudoplh Steiner, New age) and Barbara Ann Brennan.

As each body develops, thousands of possibilities open for us, depending of the kind of life we are following. Yogis say that we should focus our awareness on the soul and achieve coherence between action, word and thought. This coherence would allow for the maturation of each body. Not everybody matures all seven bodies.

Because chakra theory is the best known nowadays among those who believe there is an energy structure of the body, I offer below a few details about the chakra system.

Chakras

Chakras have been defined as energy fields or vortexes. Also, they are seen as the centers of consciousness where an exchange between the energy that surrounds us and the energy that circulates within our bodies, takes place.

What we know about chakras comes from ancient traditions as well as from people who claim to be able to perceive or intuit these energy centers. *Pendulum reading* is also used to determine some of their characteristics. I have used rose quartz pendulums to "see" how the person is distributing his energy in the body.

It is interesting to note that the concept of chakras has appeared both in Eastern as well as in Western cultures (before Internet), in various religious beliefs, spiritual practices, yoga or occultism, with slight differences among them.

In his book *Vibrational Medicine,* Richard Gerber says that "Chakras are said to be energy transformers that process a type of environmental subtle energy known as 'prana' and integrate this nutritive energy into the cellular framework of the body via thread-like connections to the major glands and organs of the body."

Chakras, according to Gerber, process different energy frequencies, translating the vibrations of the subtle bodies into physiological manifestations. Late Indian Guru Osho said each person has a different number of chakras in the body, but most authors coincide in mentioning seven major chakras with the locations and characteristics outlined in Table 3. However, there is no similar consensus about secondary or minor chakras that are most often located in palms, eyes and joints.

There is a correspondence correlation between chakras and levels of awareness, archetypal elements, developmental stages of life, vibration (colors, sounds), organs and mental and emotional functions. Five of the seven major chakras are located in the midline of the body, both in the back and the front. The first chakra is located near the end of the spine and the seventh chakra, in the crown or a few inches above.

What I offer here is a diagram, just a manmade representation of the chakra system. However, in order for you to be able to use the knowledge about chakras and energy flow in the body, or to learn about universal energy going through the reiki practitioner to a recipient, you as the learner or seeker need to go though a personal process of discovery that comes not from books but from subjective experience. Books are only the aperitif.

Throughout life, the chakras undergo a process of activation and development (awakening) in accordance with the subtle bodies. Ideally, both processes go on simultaneously.

In spiritual practices, *Ego* has been defined in different ways, many times as a foe that we must defeat as the source of all evil in ourselves. In my own understanding, the ego is a structure that operates to grant the body what it needs for survival. The three inferior chakras are the ego chakras. When we are skewed or disconnected from our soul, from the whole to which we belong, the ego takes precedence and we may be driven by fear. Thus, fear might be the source of greed, lust, gluttony, laziness, envy, wrath and pride.

Reconnecting with our essence changes our perspective, from a mortal, limited body full of needs to a creative soul that lives in an eternal present and has no hurries. The soul knows about the *law of attraction;* it knows that living in harmony with the universe's will, will ensure that all the pieces of the puzzle of our lives are always in the right place.

I see the chakra of the heart (fourth) as the bridge connecting ego and soul, through the practice of unconditional love.

Chakra 1. This center develops along the capacity of the body to provide for the basic needs of nutrition, shelter, affection and stability. The wealth of stimuli provided by the environment strengthens the person, forging a certain perception of the world. When the person becomes cognizant of available choices, he develops a sense of freedom. This chakra is related to our basic trust.

Table 3. Location and characteristics of the chakras

Chakra	Location	Color	Correspondence
The First Chakra, the Root or Base Chakra	At the Base of the Spine, near the tailbone	Red	Survival, basic physical needs, sleep, food, shelter. Reproductive system.
The Second Chakra or Sacral Chakra	Beneath the navel	Orange	The seat of sexuality, sensuality and emotions. Creativity. Genito-urinary system.
The Third Chakra, or Solar Plexus Chakra	At the solar plexus, in the centre of the trunk	Yellow	Seat of the will power, related to making decisions, taking responsibility. Digestive system. Adrenals
The Fourth Chakra, or Heart Chakra	At the centre of the chest, within the Heart	Green	Giving and receiving love, grieving for loved ones. Circulatory system. Thymus
The Fifth Chakra, or Throat Chakra	In the throat, beneath the chin	Blue	Communication, in both mundane and spiritual planes. Creativity. Respiratory System, Thyroid
The Sixth Chakra, or Third Eye Chakra	Middle of the forehead, just above the eyebrows.	Indigo or Violet	*Third Eye.* The organ of spiritual perceptions, seat of psychic visions and clairvoyance. ANS. Pituitary
The Seventh Chakra or Crown Chakra	At the top of the head	White or Violet	The spiritual door, our connection to spiritual wisdom and the transcendent. CNS. Pineal

Chakra 2. This center is also related to the process of separation and individuation and, as the former, it develops during the first few years of life. It's related to our gender identity, sexuality and sexual preference. This is the chakra of productivity, creativity and giving

birth. It is related to the process by which we seek to be as everybody else, without renouncing our individuality.

Chakra 3. Most authors relate this chakra with processing emotions. It is considered our power central and houses our identity and need for validation, appreciation and respect. It regulates the development of our own opinions from which we form criteria to make our own decisions seeking the actualization of our potential as human beings.

Chakra 4. The heart chakra develops during adolescence and connects the three inferior chakras with the three superior chakras, becoming like a gate that regulates the flow of energy. It is the center of unconditional love, impersonal love, devotion, ability to heal, and empathy. Through this chakra, we connect with others in a deep, truthful way. Through the way of the heart we can "purify" our emotions before expressing them.

Chakra 5. The throat chakra develops by the end of the teen years and the beginning of adulthood. It is the center for communication, inspiration and expression. It regulates the expression of creativity in a way that touches other people lives.

Chakra 6. Known also as the *third eye* chakra, this is the center of perception, extra sensorial perception and intuition. It governs our existential consciousness.

Chakra 7. The crown chakra is the center that connects us with what is beyond us, with the transcendent. Through this chakra, we integrate the other centers, learn to accept ourselves and to embrace our lives.

Even though the development of each chakra is correlated with a time in our life cycle, most people achieve total development only up to the third chakra.

Some authors explain initiation to reiki as the alignment or attunement of the chakras to facilitate the ascent of the Kundalini energy stored in the first chakra. For others, the attunement facilitates the channeling of universal energy through the crown chakra. The positions of the hands during reiki treatments seem to have a relationship with these centers.

As mentioned elsewhere, pendulum reading can be added to the practice of reiki as an assessment tool. There are different systems to interpret the movement of the pendulum. I think one must develop one's own system. If we use a rate chart we might end up in a *what's-wrong-with-me* paradigm, *diagnosing* ills that need to be *fixed*.

However, if we understand the movement of the pendulum as information about a particular person's distribution and use of energy in the body, then we can use the pendulum as a guide to certain courses of action. I look at the form, amplitude, regularity and direction of the circles traced by the pendulum to provide feedback that the person might use for her own process of transformation.

I prefer to see a person's individual vibrational pattern as unique and I know that the patterns vary from time to time, indicating the changes that have taken place. Many times I use the pendulum before and after the attunements to help people see how certain aspects of the pattern change after experiencing the connection with universal energy.

In addition to the seven chakras, The *Upanishads* (Vedic books written around 700 B.C.) described a network of about 72,000 nadis or subtle channels. There are 14 principal nadis, the most important of which are Ida (left of the spine, feminine, represents the moon), Pingala (right, masculine, represents the sun) and Sushumma (central channel runs from the root to the crown chakra).

❧❧❧

The whole universe is basically energy and all of its forms are controlled by information fields. The energy exists everywhere as a potential: in the conscience or spirit, in our thoughts, transmission of nervous impulses, gravity. In the body, the exchange of molecules is communication. Sound, light, heat, magnetism, color and electricity are just variations of energy. What varies is the vibrational frequency through which energy is transmitted.

The subtle bodies "soak" the dense or physical body. The etheric body *in-forms* (gives form) to each part of the physical body and establishes its functions, like a stencil determining the structure, the function and the biochemistry of the body. It corrects deviations, broadens the range of motions that favor growth and development and responds to the information coming from the dense body. At this level takes place the communication between the different bodies or dimensions, through nadis, chakras or meridians, according to the different points of view.

The chakras in one system or the acupuncture points (tsubos) in the other are the junction centers of the communication system, where the information coming from adjacent tissues is integrated with information coming from other sources, including the environment.

Blockages to the flow of energy become apparent in the physical, emotional and mental dimensions as symptoms. These energies can be detected at the subtle level by those who have developed the capacity to perceive these energies or by using special devices. Blockages can, for example, be detected as *parasite frequencies* (that do not belong there), using laser, micro currents or colors with certain devices

After discovering micro meridians in the hand that worked as mirror images of the body's meridian system, Doctor Tae Woo Yoo developed *Koryo Sooji Chim*, or *Koryo Hand Acupuncture Therapy*. He uses a small electrical device that he calls "the ray" to both assess the flow of chi through the meridians and correct excess or deficits of the flow of energy in certain areas. I had the opportunity to use one of those devices before and after a reiki treatment and I verified that the voltage at the *alarm points* (meridian points) returned back to normal after a one-hour session.

A few years ago I visited Dr Jorge Carvajal as a patient. I had been dealing with pain in both neck and shoulder. He treated me in a similar way that I have seen the pranic healers doing. At the beginning of the consultation he used his hands to assess my chakras over the aura and then he used a soft laser to do auriculotherapy (stimulation of acupuncture points and reflex zones in the ear using a laser beam). The consultation didn't last more than 20 minutes but in those few minutes, the pain that had been bothering me for at least three months disappeared right away.

The following year I had the joy to be his student at several seminars on *esoteric healing*. I learned some techniques to manipulate the energy in the body. Besides *the laying on of hands* and the laser, Doctor Carvajal was a pioneer in Colombia who introduced magnets, copper spirals and little glass "filters" to treat his patients. The *filters* contained specific tissue and molecules specimens, which when applied to certain areas of the body served the purpose of reprogramming or restoring the information that an organ needed to get back into balance.

As we see, there are several alternative practices that utilize tools to intervene the energetic stencil, changing the vibrational frequency or using cybernetic principles to treat different physical and mental conditions.

Binomial of structure and function

Most of what I learned in medical school and what guided my work as a physician in general practice for more than 17 years was focused on the components that made up the structure of the body. It was necessary to know the human organism and that required shredding it (so to speak) to the microscopic level. To understand jaundice, for example, it was fundamental to study the biochemistry of the degradation of bilirubin in the body, to observe a hepatic cell under the microscope and to dissect a liver and its ducts in anatomy and pathology labs.

I feel grateful to my docents and especially to the physiology teachers who taught me the intricate relationships between structure and function from a dynamic perspective. I also feel indebted to the teachers who in the clinical courses showed me how to analyze the bodily processes that had resulted in illness or how to interpret the lab tests in the context of clinical evidence. I extend my gratitude to masters such as Guillermo Fergusson, who was our dean and our teacher and taught us that illness had causes other than physical ones, and forced us to look at the needs of the unprivileged. With him, I understood better than ever that medicine was a service profession. Other teachers helped me see the importance of prevention and rehabilitation. All of them equipped us with a solid foundation to understand the functioning of the human body from an ecological standpoint. And where would I be without the patients who taught me that a good treatment goes far beyond prescribing medicines?

A few years ago I injured my left knee. The meniscus and the collateral ligaments were torn. I didn't accept any kind of medical intervention, convinced that even very simple yet invasive procedures like an arthrography involved risks. I wanted to believe in the intelligence of the body that I had been preaching for a while. At the beginning, I followed the clues provided by my body. If I tried to walk and the pain was intolerable, then I would stop walking. The body was indicating that the knee didn't want to bear my weight and I had to accept that the pain aimed to prevent me from bending the knee, which could cause further damage. I got a knee brace so I could move without pain, drive and go to work. It was not necessary to take pain killers or antiinflammatories, which are a first-line prescription for all injuries. I knew that numbing the pain was cutting the communication between my body and my mind. I used reiki and magnets, and the pain and inflammation were minimal.

My main preoccupation became to avoid limiting sequels. After only three weeks, I had restarted my morning walks. I was so enthusiastic about my fast recovery that I went to a street Latin jazz festival in Miami where I spent most of the morning walking from stage to stage, eager to get the most of the event. I paid for my negligence. I re-injured the knee. Now pain and inflammation were really serious. The knee was shouting to me, "Learn to accept your limitations!" The body made me rest. I used reiki, magnets, visualization and Trager (a type of neuromuscular reeducation therapy). The pain receded but I hadn't yet learned enough and I injured the knee once more when I tried to force a flexion. My body was now shouting really loud. Reiki, magnets and Trager again helped me, but now I was ready to listen. Now I could understand, with a depth that I have not achieved before, the relationship between structure (my bones, cartilages, ligaments) and function (performance).

I also explored and understood the symbolic meanings of the knee symptoms.

Among the few books that I always keep close to my bed is *The Healing Power of Illness*, written by a psychologist, Thorwald Dethlefsen, with a physician, Rodiger Dahlke.

I've carried it with me for years, since I was introduced to the idea of the symbolic meaning of symptoms. This is not the kind of book I'd devour from cover to cover, and I don't agree with everything it says. It has, rather, been a book that I personally use to listen to my own intuitive answers while pretending that the knowledge comes from an outer source. (Isn't it interesting that we tend to trust external sources better?)

Dethlefsen and Dahlke depart from conventional diagnosis-making. It was from them that I learned, for example, that extremely responsible people suffer colds mostly when they have a need to slow down, isolate from others and take a break without feeling guilty for thinking about themselves. Or that diarrhea and constipation are physical symbols informing us about our fears and need to let go of things.

According to the German authors, "Symptoms make us honest. In our symptoms we have what our consciousness lacks." For them, physical symptoms are the expression of conflicts in other dimensions of our being that we have not yet acknowledged.

To understand what illness is about, its purpose, you start by asking which aspect of your living is represented in the nature of the affected organ. Kidneys represent partnership; the skin represents the way we relate to the world (contact); shoulders bear the weight of our burdens and so on.

The second step is to determine what kind of illness it is. Inflammation (redness, swelling) usually represents anger, for example. And, for step three, the authors take you system by system, supporting the symbolic interpretation of symptoms with popular expressions. What is most interesting is that the book was originally written in German and the translated expressions are equivalent in English and Spanish. "What are you unable to swallow?" would be the question for a sore throat. "What is painful to hear?" would give you the clue for your earache. "What's eating you?" is the question for a gastric ulcer.

According to the authors, symptoms such as high blood pressure could be representing a psychological blockage that represses the energy a person has put into achieving a goal.

Let me give you an example of how I have applied what I learned about the symbolic meaning of symptoms (and don't just believe me, but put it to a test). In many occasions when I hear a close friend saying, "It seems that I'm getting a cold," I ask them a simple yet irreverent question: "What for?" They make me repeat the question because they think they heard me wrong, but I patiently bear with their suspiciously sarcastic smile, explain the possible purpose of the cold and imply that it would be easier to recognize the conflict and the need, instead of falling ill. Friends seldom agree with me (openly at least) but somehow, after they have thought about it, they never get the cold. Want to give it a try?

Louise Hay gives similar clues in her book *Heal Your Body*. And in *Your Body Speaks Your Mind,* Debbie Shappiro offers explanations of the same sort. Fifteen years ago, the words of these authors belonged to a special niche: alternative medicine. Today, science has advanced much in the direction of establishing the unity of body and mind. Or as Candace Pert puts it, that *the body IS the mind*.

Science has proven that our emotions trigger the release in the body of certain molecules that carry orders from *command central* to the organs in charge of preparing the body to adapt to external challenges.

However, we cannot establish with the same certainty how an emotion directly correlates to an alarm rung by a specific organ. But we know that our brain cortex is connected with the deepest layers of the brain, where an important part of our *inner healer* directs the process of renovation, restoration and repair of injured tissues. And we also know that the right and left hemispheres of the brain perform different functions; the right one being the one in charge of symbolic meanings.

In *The Body Believes Every Word You Say*, Barbara Levine lists examples of words we unconsciously use every day and how these words may plant a seed for illness. She went through the personal experience of a life-threatening brain tumor and began studying how the words she used reinforced her condition. Her book is the product of 15 years of research. She suggests, for example, that expressions like "that breaks my heart" can set us up for a heart attack.

Nevertheless, it doesn't mean we are to blame for our illnesses. Read this paragraph from Shapiro's article, *Curing a Symptom or Healing a Life*:

"Through illness the body is giving us a message, telling us that something is out of balance. This is not a punishment for bad behavior; rather it is nature's way of creating equilibrium. By listening to the message we have a chance to contribute to our own healing, to participate with our body in bringing us back to a state of wholeness and balance. So, rather than blaming ourselves by saying 'Why did I choose to have cancer?' we can ask 'How am I choosing to use this cancer?' For we can use whatever difficulties we are confronted with in order to learn and grow, to release old patterns of negativity, to deepen compassion, forgiveness and insight. Our difficulties can then become stepping stones along the way rather than stumbling blocks. Instead of becoming overwhelmed by a sense of hopelessness and guilt that we are responsible for everything that is happening to us, illness can be seen as a tremendous challenge and opportunity for awakening. In this way, illness is a great gift - a chance for us to find ourselves."

When I injured my knee, I had been initiated to the third level of reiki recently and I felt urged to learn humility; however, I mistakenly took a wrong path, which led me to accept humiliation at work. The hint that I was in error was provided by my knee that, because of the pain, didn't allow me to kneel down. At the same time that I accepted my limitations I also overcame pride and became more assertive. I learned patience and benevolence, accepting that, in order to heal, the tissues required time and that only when the tissues had healed would the function be completely restored. I introduced movement and stretching bit by bit to overcome restrictive residual patterns caused by the pain and after six months my knee had reestablished total function. As of today, the knee doesn't bother me at all.

In these dualities–function/structure, physiology/anatomy, freedom/limitation–I understood the interplay between unified opposites that complement each other at the same time they contradict each other, which impels change, and, in this particular case, healing.

Within a lineal framework, attention is focused only on one of the poles of a unity and not in the unity of contraries to the point that one of the elements of unified opposites is denied or overlooked. Cause and effect are also a unity of opposites where one becomes the other, depending on the circumstances.

When I speak about structure I refer to the parts of which our bodies are made. When I talk about function I mean process. Structure comprises the different levels in which we exist, from energy, to genes, to molecules, to systems. Medicine is strongly interested in the relationship between structure and function. It is a dialectic principle that we understand better thanks to Darwinian theories that establish how function determines structure and vice versa.

However, in this stage of development, medical research and treatments are focused on modifying the structure, be it with surgical interventions, bioengineering or medications. Even if it is true that these treatments aim at recovering or improving function of the affected part, in most occasions, they intrude on the body in such a way that they disturb functionality. A good example is the stomach bypass that is performed to treat diabetes in obese people.

Back to the knee, I needed to splint the joint, accepting the need to limit its function, to prevent further damage to the structure. When the pain became tolerable, I should have tested the knee little by little to give the body time to regenerate and repair tissues. I should have swung between rest and movement to allow the body to assess possibilities of movement and actual limitations. But when I exceeded the limits, the muscles around my knee created a shield of spasm that further limited my movement. When I tried to force this contracted muscles to move I prompted an alarm – pain – that extended the spasm. It was necessary to reeducate muscles and nervous system using Trager to help the nervous system monitor, with the help of a therapist, the boundaries within which movement was safe, so that muscles could relax where they were still tight because my mind had added fear to the picture. It was also important to keep moving within a certain range so the body was capable of realizing the function that was possible based on the information provided.

The musculoskeletal system is made up of bones, joints, tendons, ligaments, cartilages and muscles. Muscles are flexible and are capable of contracting and relaxing to perform their work. A cartilage within a joint is a little softer than the bone it covers, with a smooth surface that has no edges, so that it facilitates the sliding of one bone head over the other and thus it grants movement and locomotion. The structure is designed according to the function.

What happens when a human being becomes sedentary? All of us have heard about the risk of osteopenia and osteoporosis (loss of calcium in the bones) that appears both when we overexert ourselves or when we forsake physical activity. We have also heard about sarcopenia (loosing muscle mass) that appears after age 50 as a consequence of a sedentary life. These are a couple of good examples of how the function affects the structure. An unused organ tends to atrophy or shrink.

This relationship between structure and function can also be seen from an evolutionary perspective that explains the natural nooks and crannies of our bodies. For example, over centuries, human beings with the most functional skeletons (that allowed them a better range of movement) succeeded in their fight for survival, enrichening the species' gene pool with individuals that have a skeleton that facilitates running (to escape from danger). A similar process has occurred to each organ, with each structure of the human body. Natural selection explains the *Homo Sapiens's* body, its structures and corresponding functions.

I won't argue about the need for surgery to modify structure in cases where the life of the person depends on the intervention. There are also reconstructive surgeries that ensure the recovery of functionality and a return to work. Nonetheless, statistics show that surgeons are performing thousands of prophylactic surgeries that could be avoided if patients made a commitment to adopt healthier lifestyles.

Hysterectomies are frequently performed "to prevent cancer" even when it seems unlikely that it will ever occur. Many doctors argue that once the patient has surpassed the reproductive age she doesn't need the uterus anymore, even when it's well known that fibroids (the most common reason to recommend the surgery) usually disappear after menopause. More examples: U.S. statistics show that hundreds of unnecessary cesarean sections are performed each year. Many cosmetic surgeries have ended in adverse events or undesired results. Many amputations could have been prevented.

A 1999 report released by the *Institute of Medicine* (IOM) entitled *To Err is Human*, established that between 50,000 and 100,000 deaths in the United States could be attributed to medical errors. Based on a study of 37 million patient records, the healthcare quality company *HealthGrades* released another report in 2004 affirming that "an average of 195,000 people in the country died due to potentially preventable, in-hospital medical errors in each of the years 2000, 2001 and 2002."

The fourth cause of illness is secondary effects of prescription drugs. Pharmacologic and surgical solutions are used in excess due to a mentality where it seems preferable to intoxicate the body with chemicals, notwithstanding the side effects, and a resistance to see illness processes from a more dynamic, ecological perspective.

It's clear that many of the medical errors are due to ill design of hospital systems, physical exhaustion on the part of doctors and nurses, excessive responsibility, fear of malpractice suits and other pressures that generally fall on the shoulders of health care professionals in the hospitals.

For many years, conventional medicine has focused on combating disease using "antis:" anti-inflammatory, anti-histamine, anti-biotic, and interventions on the structure including radical surgeries, amputations, and joint replacements. Notwithstanding, it seems hopeful that from the industry of illness we seem to be moving towards the industry of wellness.

Magazines publishing information that can help people lead a healthier life and seminars training people on how to restore inner balance are in the hundreds, and natural products (supplements, vitamins, minerals) that can boost health are easily available in health stores.

What will come out of this? New dependencies from new products, or an increased awareness? As technology has invaded us and changed our lives, it has to provide us with solutions, claim the companies that sell natural products. We may look with optimism on the resurgence of a medicine that is more interested in stimulating functionality, a medicine that is looking for ways of preserving the structure while optimizing its functionality.

The body wisdom

In this section I will talk about the systems of the body, summarizing their functions from dynamic, ecological and holistic perspectives. My purpose is showing the *inner healer* in action. It will be just a panoramic view on how all the organs get engaged in the self-regulating, self-regenerative and self-healing processes of the body.

Let's look at an example of cooperation among systems first:

The lungs inhale air and take oxygen from it. In the lungs' tiny alveoli, red blood cells (RBC) receive the oxygen and transport it to the tissues. For that purpose, RBCs use a molecule known as hemoglobin, partly made of iron.

This iron comes from what we eat; the digestive system must remove this element from the food. The circulatory system takes iron to the liver for storage or directly to the bone marrow where the hemoglobin factories are located. Bone production rate is regulated by thyroid hormones (Calcitonin) and parathyroid hormones (PTH) and it is also stimulated by muscle contraction. Sugar produces the necessary energy to make the muscle contract. Sugar, coming from our food or from our body storage, requires insulin to enter the cells in order to be used. Insulin is produced in the pancreas. Other hormones produced in the same organ regulate the amount of insulin the pancreas secretes.

The above is a good example of systems interaction and interrelation. In order to perform their tasks, the different organs require an optimal communication system.

Scientists have not reach consensus on some of the studies I present here. Nevertheless, in science there is never unanimity, there are always questions. Thus, if the most daring ideas I present here do not convince anyone, I will be happy if at least they raise some eyebrows – if not questions.

I deliberately avoid talking about pathology, although sometimes I mention certain conditions when I feel it's necessary to make my point. From a holistic view, I would affirm that *illness as such does not exist*. As we have discussed, the body is constantly adapting to environmental changes that it perceives as stressors. Adaptation is granted thanks to body responses that sometimes fail in keeping the necessary equilibrium either because of the intensity and extension of the stressor or because the capacity to respond has been weakened.

To study the human body we have no other option but to classify it in body parts, systems or apparatus. But we must not lose the sight of the whole and of the interconnectedness of the different parts of the body. As you read, please keep in mind that even though we are talking here mostly about the physical aspects of the body, it is multidimensional and thus, the adaptation processes have their expression in the physical, the emotional, the mental and the spiritual levels. The different responses of the body to different stimuli involve responses in both the dense and the subtle dimensions of the body as well.

I. Tegumentary System: Our shield

Our skin constitutes a boundary between our dense body and the external world. Being impermeable, it ensures the stability of our inner environment.

The skin produces a protective mantle, which keeps it lubricated preventing dryness, counteracting the effect of toxic substances in the air and water and keeping at bay the population of bacteria that live on our surface, Moreover, thanks to the presence of *elastin* and *collagen*, our skin is flexible, allowing movement.

When exposed to the sun, the skin produces *melanin* that gives it a brown hue, preventing the harmful effects of ultraviolet rays from the sun. Melatonin does not block the rays as a sun-blocker does, though; the body permits ultraviolet rays because they are necessary for the production of Vitamin D, which promotes the formation of bone tissue.

The skin constantly monitors the environment, sensing changes in temperature, detecting obstacles and alerting about possible injurious stimuli. The nervous system responds to the signals sent by the skin by communicating with other *interested parties* in the body and creating an *action team* that will respond to changes or threats.

When I am touched in the middle of my back, I know the exact point where I was touched, without looking. The information sent by the sensors in the skin helps the brain configure a map that the body uses as if it were a *Global Positioning System* (GPS). The brain communicates with the skin for a full assessment of the situation. "What kind of object touched me? Is it hard, warm, cold or dangerous?" Based on the answers, the brain continues to enquire. "Do I like it or not? Do I allow it or avoid it?

Touch defines us in relation to what surround us. While sliding through the birth canal we receive a massive tactile stimulus that we never experienced while we were floating inside the womb. This experience helps the brain create the first map of the body, which will be redefined once and again as we grow up, develop motor skills and experience the changing environmental conditions. Touch and movement are essential for the development and updating of the map.

The skin also plays a role in the control of body temperature. It is an essential role because most chemical reactions in the body require certain ranges of temperatures. If the body raises its temperature (after exercising, for example), then the blood vessels under the skin dilate, the skin irradiates heat and it sweats to keep the temperature stable. This process is mediated by messengers sent to the sweat glands and the blood vessels by the brain.

If the temperature goes down as the result of the body being exposed to extreme cold, then the blood vessels narrow and the heat is preserved in the vital organs inside the body.

Sweat not only serves the purpose of helping to control temperature, but also to excrete toxic substances.

II. Connective System: wrapping and binding the organs

There are various types of connective tissue. The name derives from its function, because this is a system that connects organs with organs: muscles with muscles, muscles with tendons, tendons with bones and so on. This communication system is made up of bones, cartilages, fat and blood. Its functions include support, transportation and protection.

Here I will give special attention to the fascia, a sheet of fibrous connective tissue separating and binding different structures of the body. It's a continuous sheath that in some places is fairly thin and in others quite dense; here it is superficial, there it's really deep. It is found just below the skin, where it's known as hypodermis, or enveloping muscles and bones, the heart and the bowels, the lungs, and all of the other bodily organs.

Biologist James Oschman postulated that the fascia penetrates even individual cells, connecting them with the cytoskeleton[31].

He also has suggested that the network of connective tissue that holds together all the parts of the body have a correspondence with the horizontal ramifications of the acupuncture meridians (called luo).

The Hungarian scientist Albert Szent–Gyorgyi, pioneer of *Electronic biology,* felt that biological responses were too fast to be totally explained by the synaptic connections of the nervous system or the ordinary biochemical reactions taking place in the body. He formulated the hypothesis that mobile electrons would be a conduit for energy and information from one place to another in the organism. But in order to perform this, the electrons needed a conductor. In 1941, he suggested that the proteins in the body were such conductors. Now science knows that all proteins in the body actually perform as semiconductors,[32] carrying both electrical currents and information. This means that the fascia, as a matrix rich in collagen (a protein), is a semiconductor network.

In other words, we are wrapped inside by a continuous tissue that comprises an electronic communication network with the ability to detect and conduct energy and information and to store and process the latter. Such a network grants unity to our structure and functionality to our organs.

According to Oschman, this system would respond in a specific and sensitive way to sounds, light frequencies, magnetic fields and physical contact.

Physician Richard Gerber, in his book *Vibrational Medicine,* reported studies done by Kim Bong et al. in the seventies, in Korea. They explored the nature of the meridian system in animals and found that there was a correspondence between the location of meridians and a fine system of tubules located under the skin that are independent from the circulatory or the nervous system.

---

[31] A network of fibers throughout the cell's cytoplasm that helps the cell maintain its shape. (Source: Biology.about.com)

[32] Materials that have both electrical conductivity and insulation properties.

These studies were later verified by Pierre de Vernejoul who in 1988, with his colleagues, injected radioactive isotopes into humans and found that the substances moved along the meridian tracks.

Hong-Qin Yang et al. presented a study with 30 subjects to the *International Symposium on Biophotonics, Nanophotonics and Metamaterials* in 2006. Using infrared thermal imaging they found evidence for the objective existence of acupuncture meridians structure in human body.

Because of the current flowing through the superficial fascia, or the meridian system, some authors have suggested that both fascia and meridians are an interface between matter and energy, between the dense and the subtle bodies.

Janet Travell, whose work with David Simons merited her to become the White House physician during John F. Kennedy's presidency, designed in the forties a neuromuscular therapy that is used by many therapists today. The modality is based in the treatment of *trigger points*, little areas of extreme muscular tension that refer pain to other places in the body. It is not unusual for the pain to be more intense far from the point of origin where an examination will show no anomaly. The referred pain cannot be explained by the existence of a neural pathway[33], but tracking down the pain to its origin traces a very similar line to that of a meridian.

According to Travell, most of our muscular pain could be caused by *trigger points*. In fibromialgia and chronic fatigue, trigger points have been localized in very specific points of the body. They explain the pain but not its cause. These points can be felt as knots under the skin and they emit specific electrical signals that can be measured with proper equipment. The constant tension in these points seems to affect the fascia that envelops the muscles, and would explain at least in part the referral of the pain.

The characteristics of the fascia – position, tone – make our bodies unique. Fascia is responsible for our body's shape.

Many of us have had the experience of, for example, a sprained ankle that is followed by certain soreness in the knee and then maybe by some pain in the lower back, just after the ankle feels better. We might not see the connection between the back pain and the injury of the ankle; however, if we trace it back we can realize that the strained ankle made us move in certain ways to protect the injured area

---

[33] Tracks that connect one part of the nervous system with another.

(compensation, in medical terms), with muscles playing in a way they haven't done before. Without our awareness, we have modified our posture and this has caused the subsequent symptoms. When we regain balance, we can break the restrictive patterns generated first by the pain and subsequently by our fear of hurting again.

What extends the pain farther than its protective function? Why will a fracture of the ankle cause back pain five years later? Is it possible that part of the explanation resides in the tensions of the fascia that the body has not been able to resolve? (Or even in the mind that keeps the tension in the fascia out of fear of injuring the ankle again?)

The fibrous material the fascia is made of has similar properties to gelatin. If you rub it or warm it up, the bonds that keep the collagen fibers together break down and it liquefies. That explains in part why a massage helps to dissolve tension. Therapists who practice Rolfing[34] explain that friction contributes to the reorganization of the taut fibers that are generating pain.

Fascia suffers changes in its texture and resiliency according to the way we live, our response to stress, physical activity, postural and movement patterns, and nutrition. Could changes in the fascia affect the inner communication of our body; cause energy blockages?

III. Musculoskeletal System: structure, movement and connection

Without locomotion, how would we be able to interact with nature? Isolation in a restricted area that limits our movement is one of the worst punishments. Full participation in life is difficult for a person who suffers from a physical handicap that restricts movement.

Muscles and bones not only contribute to shaping our physical body, they also facilitate interconnectedness with other living creatures.

Evidently, the whole body participates in some degree to make locomotion possible. Lungs take oxygen that is transported by the blood to the cells so that the oxygen *burns* the sugar that we got from food or from our body's *warehouses* to produce the energy that the muscles use in a contraction. The nervous system assesses posture,

---

[34] Rolfing is a type of therapy based in the manipulation of fascia and muscles. It looks to restore *alignment* of the body to improve function and physical and emotional health.

distances, and existing obstacles and then delivers messages to the muscles to coordinate movement.

Movement is the modus operandi of the universe, as Christine Caldwell[35], founder of the *Department of Somatic Psychology* in Naropa University, CO, puts it. Everything that is alive moves, and life needs a structure to perform movement.

Caldwell notes how in the locomotive system the relationship between structure and function becomes quite obvious. Wings are needed to fly, fins are needed to swim, and bipeds and quadrupeds need legs to walk, run and jump. When, in the process of evolution, Man became *Homo erectus,* the legs got stronger to make it possible to carry the weight of the body and to free the hands for other purposes.

Movement is also regulated by *feedback loops.* There is a constant communication between the nervous and the locomotive systems. Muscles, tendons and joints have neural receptors that send signals to the body to indicate a special position, degree of tension in a muscle or tendon, or the pressure exerted on a joint.

We are never totally relaxed, not even in our sleep. To keep a certain posture, groups of muscles take shifts. When joints feel the pressure caused by muscle contraction and gravity for a time, they signal another group of muscles to take over. While we are moving, if a muscle is contracted its opposite (antagonist) has to relax to grant movement.

Muscles, joints and skeleton also participate in body language. Posture, gestures and signs express our mood, needs and affections. Scientists have found that facial expression is not only trans-cultural, but the same group of muscles contract in humans and animals while they are experiencing similar emotions.

The organs that make up the locomotive system play a crucial role in the maintenance of stable conditions within the organism. Muscles contribute to keep temperature stable, which becomes evident when we sweat after physical exercise or the temperature raises after we have experienced fever chills.

Our bones are made up of two different kinds of cells, *osteoblasts* and *osteoclasts.* The former build, the latter tear down bone tissue and by doing so they help to keep stable amounts of calcium in the blood.

---

[35] From a paper Caldwell shared with me entitled *Life Dancing Itself: The role of movement in play and evolution.*

These cells also help the bones change form and density according to challenges posed by gravity.

As we have mentioned, walking is healthy not only because muscle contractions benefit circulation, but because it promotes bone building, muscle tone and flexibility, not to mention that it supports respiration, stimulates the nervous and the immune system and the production of endorphins.

Moving our legs tells our body that we are alive. The function *informs* the structure; the structure is defined according to the function.

Movement patterns develop and evolve from the moment we are born until we die. Every life event leaves footprints on our framework.

IV. Circulatory System: transport and perfusion

Every single system in the body depends to a greater or lesser degree on the circulatory system. But this on turn depends on the other systems to perform optimally.

The circulatory system is the transportation system par excellence. Hormones, antibodies, cellular byproducts, oxygen, all of them depend on this system to arrive at their destination and play their part in the task of keeping the body functioning.

Blood circulation depends on the constant pumping of the heart, on the large vessels that dilate and contract and branch once and again until they become microscopic and deliver their *cargo* of nutrients and oxygen to the cells.

Plasma, which becomes lymph, is filtered from the blood towards the intercellular spaces through capillary walls. Much of this fluid is either absorbed by the cells or returned to the bloodstream, while a small amount remains in the vicinity of the cells to keep a reserve supply of nutrients available. Excess fluids drain back through the lymphatic capillaries into the venous blood, a little before the stream reaches the heart.

Lymphatic vessels also transport proteins, fats and other substances that either escaped from blood vessels or are a directly byproduct of cell activity.

Lymph flows even against gravity thanks to breathing and muscle contraction. When we inhale, a negative pressure is created in the chest that sucks the lymph and facilitates its circulation. When we exercise, the muscles flush the lymphatic vessels, acting like a pump.

We would not be able to excrete the residues of cellular activity without the circulatory system. Many of the functions of the immune and endocrine systems are facilitated by the circulatory system, which makes it possible for the different organs to communicate as the blood makes its run from the heart to the tiniest of capillaries.

The circulatory system is partly self-regulated and partly regulated by the nervous system, which guarantees the process of distribution and redistribution of the blood, according to the needs presented by the cellular activity and the adaptive needs presented by the environment. The system guarantees a stable flow pressure as well as the needed amount of fluids in the organism, redistributing them continually, according to the demands of the body.

Priorities are set by the *inner healer* in such a way that the blood will rush to the areas of the body that need a higher supply during performance. Just remember parents repeating over and over that, "You can't swim after you eat." After a meal, the digestive system draws more blood than usual from around the body to help with the digestive process. There might be less blood, and thus less oxygen and nutrients, for muscles and the brain, increasing the chances of cramps and retarded responses. This also explains why people get drowsy after a meal.

The intelligence of the body in this case is manifested both in the wise distribution of the blood and in the maintenance of an even heart pace and blood pressure to grant the flow.

Both the sympathetic and parasympathetic branches of the nervous system account for the frequency of heart contractions. There are also feedback loops and reflexes in charge of this regulatory process.

V. Respiratory System: O_2 for internal combustion

Energy is required to keep all bodily systems working properly. For the production of energy, a carburant is necessary and in the case of the body the carburant is the oxygen that we get from the air as we breathe. On the other hand, carbon dioxide (CO_2) the byproduct of cellular work, needs to be expelled from the body. Therefore, the body is set up for a gas exchange, oxygen for CO_2, and this happens at the level of small sacks called the alveoli in the lungs.

We call this exchange of gases in the respiratory system pulmonary respiration. We inhale oxygen and we exhale CO_2.

We call the process by which oxygen is transformed into energy, at the mitochondria in the cells, cellular respiration.

The respiratory system cannot operate well without the proper functioning of the circulatory system, which is in charge of picking up and distributing the oxygen from the lungs to the cells and taking CO_2 from the cells to the lungs to be expelled.

The number of red blood cells the body needs to produce for efficient transportation of oxygen depends on many factors. For example, if the concentration of oxygen in the atmosphere is low, such as in places at high altitudes, like Bogota, the body knows that it needs to increase the production of these cells. The opposite happens at sea levels.

Each element of the respiratory system plays a certain regulatory role. The nose keeps the air humid and warm, to prevent it from damaging the air pipes. We see this when we are exposed to a low temperature and the nose becomes runny, producing a watery secretion and we sometimes think that we've caught a cold. However, it's just the body responding to a change in our environment.

The bronchi are lined with *hairy* cells that brush away any particulate matter that we have breathed in. The sinuses contribute to the production of mucous that, with nasal and bronchial mucous, trap tiny objects that could irritate the system.

Some schools of thought have given special attention to the way we breathe. Yogis, for example, teach breathing techniques to students. Deepak Chopra, in his book *Ageless Body, Timeless Mind: The Quantum Alternative to Growing Old*, says that breathing intimately connects us with the universe and articulates body and spirit. For him, each cell is a tiny terminal connected with the *cosmic computer*.

In TCM, respiration is also an instrument that adjusts the energy flow in the body. According to the Chinese, energy accumulates in the lungs and is then pumped through the lung meridian through the body.

Yogis believe that by controlling the breathing they can affect the heart function, which science corroborates. Deep breathing techniques lower the blood pressure and the heart rate and have a calming effect.

Certain therapies – such as *craniosacral* and *bodytalk* – assert that the body *beats* in sync with breathing. In the latter, the idea is that every time we breathe the brain scans the vibratory frequencies of the body, which allows the body to regain balance by correcting *parasite* frequencies.

It is easy to see how our emotions manifest in the way we breathe. There is a great difference in the breathing patterns of a person who is scared and another who is serene. It is also easy to corroborate that changing the way we take in the air changes our mood. Therefore, the *breathe-deeply* advice is provided to a person who is in shock or must relax.

The respiratory system also plays an important role in the maintenance of the acid-base balance in the body. Most residues produced in the 60 trillion cells (60,000,000,000,000,000,000!) of our organism are of an acid nature. Acids are neutralized by alkaline, and the blood maintains a certain excess of alkaline to balance off the acid production in the cells.

On certain occasions, like when we overexert or after a prolonged fasting period, this balance can be disturbed. Lungs and kidneys kick in; our breathing rhythm changes and our kidneys excrete excess acid or alkaline ions, depending on the case.

It is interesting to note that in Chinese Medicine Lung and Kidney meridians are interrelated.

A respiratory disturbance could also cause an acid-base imbalance. Have you ever observed a person who is hyperventilating? During panic attacks, breathing becomes deep and fast and, if this pattern persists for a while, carbon dioxide levels drop, creating an excess of alkaline that will produce dizziness, bloating, dryness of the mouth and even weakness, confusion and sensation of pins and needles in the limbs. Sometimes it can also cause seizure-like spasms of the hands and feet, chest pain and palpitations. Hyperventilation is also used for therapeutic purposes in approaches such as *Holotropic Breathing.*

In the opposite scenario, when the person has a lung condition and ventilation is restricted, carbon dioxide accumulates in the body. The extra amounts of CO_2 molecules combine with water to form carbonic acid, which contributes to increase acidity (pH) in the body.

One of the most important homeostatic challenges for the body is to maintain acid-base equilibrium. Our survival depends on it. Even a slight acidification of the blood generates significant changes in our cellular metabolism and the weakening of the immune system.

Alkalizing diets (like the one I follow) seek to support the functioning of the immune system. These diets are based on an increased intake of fruits and vegetables, and less intake of meats and dairy products. Fermentation caused by mixing certain foods is also avoided.

Organs in charge of producing sounds are also part of the respiratory system. Recent research shows that vibrations produced in the palate, cheeks and nose when we hum (say, repeating mantras like OM), improves the exchange of gases in the lungs and strengthens blood circulation. Researchers wonder if this phenomenon explains the greater mental clarity experienced by people who repeat mantras or hum while they meditate.

VI. Digestive System: Recycling, transformation, disposal and storage

Why do we have the need to nourish ourselves? An almost painful sensation in the stomach informs us that the time to eat has come. Nature makes sure that we obtain a sufficient supply of nutrients using mechanisms such as an alarm for hunger, selective appetite – which unfortunately our compulsive eating habits are annihilating – and granting us pleasure derived from eating.

Hunger and postprandial satisfaction involve the nervous system as well as messenger substances.

For years it was thought and accepted that hunger was a response to lower levels of blood sugar, which would drop after a few hours of fasting. More recent research sees hunger as related to the need of the body to keep stocks of fat at constant levels. Other researchers have found that hunger sensations are prompted by our intention to eat, which makes the pancreas secrete a shot of insulin. By diminishing the levels of blood sugar, we will feel hungry. Maybe the real explanation is a summation of different theories.

In the digestive system we again see the complex communication mechanisms that operate in the body to stimulate its functioning. Sight and smell induce salivation that prepares us to digest the food. Mechanical events such as the distension of the stomach and the intestines; chemical events, such as the presence of fat or acid in the food, and even emotional events, such as getting upset while we eat, disturb not only our appetite but also our capacity to ingest, digest and eliminate food.

The digestive system is a field where we easily see the effect of our emotions on the body. A couple of examples: people with type A personalities[36] suffer more frequently from ulcers and people given to postponing wrath expressions suffer frequently from constipation.

Let's note the correlation between the energy structure and the digestive system. The solar plexus or third chakra, located between the navel and the ribs, is said to regulate both the digestive system and the emotions.

A good digestion requires good mastication, so that by chewing, food is finely broken down and thoroughly mixed with enzymes present in the saliva. This facilitates the transformation of the foods and the best use of the nutrients contained in them. Remember that these nutrients will become our tissues (*we are what we eat*). Both conscientious chewing and a good amount of enzymes prevent excessive fermentation in the digestive track that would cause bloating, cramps and *dis-ease*.

Several organs belonging to the digestive system also participate in functions other than the transformation and assimilation of food. The liver, for instance, which has a very important digestive role (excretes bile to help in the digestion of fats) and is the largest gland in the body, works closely with almost every other system in the body. It is like a factory where many substances undergo a makeover. Some substances are converted by the liver into simpler molecules so that they can be stored in the body, and surplus substances that cannot be stored are converted into waste products to be eliminated by the kidney. The liver is also an important warehouse for glycogen, a sugar precursor, and responds to the energetic needs of the body by releasing sugar into the bloodstream or generating heat. On top of it all, the liver is a detoxification center. How does the liver know about the needs of the body?

The pancreas produces important enzymes that contribute to the transformation of fats and proteins in the bowels and it is also an endocrine gland that secretes insulin and glucagon, which the body uses to regulate the levels of sugar available to cells.

The stomach also has some endocrine functions. It secretes *gastrin* that regulates motility, and other essential substances for the digestion of food.

After we eat, the food is propelled through the digestive track thanks to a wavelike movement called *peristalsis*, which is regulated in part by the autonomic nervous system. Receptors in different parts of the system are sensitive to different stimuli. In the upper part of the

---

[36] People with a tendency to be driven, strict, perfectionist, rigid and impatient.

small intestine, the presence of fat stimulates the production of substances that slow the intestinal transit to guarantee enough time for the saponification and absorption of fats. If the mucous membrane of this part of the intestines feels the presence of acid or distends as the food passes by, the autonomous nervous system receptors are stimulated and respond by ordering the slowing down of peristalsis.

A great deal of what we eat is absorbed in the small intestine, where cellular mechanisms interplay at the level of the membrane, namely osmosis, diffusion and active transport, according to the need to absorb sodium, sugar or fats, respectively.

When food waste arrives in the colon, other receptors are stimulated and these command the intestine to accelerate peristalsis so that the residues are eliminated.

When I go to health food stores I see that *liver detox* and *colon cleansing* products occupy quite a bit of the shelves. Practitioners who advocate for liver detox, colon cleansing medications, colonics and the like, explain that the constipation associated with consumption of processed foods facilitates the absorption of toxins through the colon into the blood stream and through the liver.

The functions of the colon include absorption of water, regulation of electrolytes and, to a lesser degree, absorption of undigested food. The regulation of water and electrolyte transport in the colon also involves the complex interplay between endocrine and neuronal pathways.

In Ayurveda, Hindu traditional medicine, treatments start with fasting, enemas and laxatives to contribute to a detoxification process.

As a physician, I witnessed the side effects of the use of laxatives and enemas and so I am not inclined to recommend them. Most laxatives, for example, affect the consistency of the stools because they irritate the mucous membranes of the intestines, which might bring serious complications to people who suffer from conditions such as diverticulitis.

Personally, I prefer light fasting, where I drink only herbal teas with honey and bee pollen for no more than 24 hours. If I want a laxative effect, I increase the intake of fruit, raw vegetables and water and I eat less protein (protein seems to contribute to constipation). Eating pineapple before breakfast, without mixing it with other foods, seems to cleanse my system quite well.

VII. Excretory System: Waste disposal, detox

Not everything that we ingest or breathe is necessary or even good for the body and the body has to find a way to discard toxins and surplus. Each one of our cells needs to get rid of the waste generated during the production of proteins and other substances and for that purpose the cell has its own excretory system. The waste crosses the cellular membrane and reaches the fluids in which the cell is embedded and is finally carried away by the circulatory system. Through the lymph and the blood, waste is transported towards the urinary and respiratory systems and to the skin, to be eliminated.

The urinary system plays its part by maintaining a constant balance of fluids and electrolytes within the organism and by excreting the nitrogen byproducts of protein metabolism. The kidney is very selective in the process of excreting waste.

By *screening* the blood and *sensing* its pressure, the kidney *knows* which amount of certain substances must be discarded or retained and does so by filtering the blood through the *glomerular* membranes. The kidney keeps constant the levels of electrolytes, urea and water in the body by filtering, secreting or reabsorbing sodium, potassium, chlorine, urea and water, for optimal functioning.

Molecular messengers also play an important role in this system. Among the messengers provided by the endocrine system are *aldosterone* that is secreted in the adrenals and the *antidiuretic* hormone, secreted in the pituitary.[37]

The urinary system offers another precious example of how the *inner healer* works. If, because of dehydration or a hemorrhage, the body loses circulating fluids, the blood pressure will immediately drop, which is perceived by the hypothalamus in the brain as an alarm ringing. The hypothalamus commands the pituitary to release antidiuretic hormone right away. The kidney receives the message and responds by increasing the reabsorption of water at the level of tiny tubules and by decreasing the urine filtering rate, which prevents a further drop in blood pressure.

On the other hand, the drop in blood pressure is also detected by tiny receptors located nearby the filtering system units (glomerular membranes) that make the kidney secrete *renin*, another messenger substance. Renin stimulates the production of *angiotensin*, which commands the adrenals to produce aldosterone, which in turn tells the

---

[37] Antidiuretic refers to the inhibition of diuresis and diuresis refers to urine excretion.

kidney that it is necessary to reabsorb sodium to contribute to water retention and replacement of the lost fluids.

VIII. Reproductive System: perpetuating life

The main function of the reproductive system is to ensure survival of the species. The organs that make up the male reproductive system create the conditions needed to guarantee the presence and viability of the sperm on the way to the female ovule so that fertilization can occur. The female reproductive organs, in turn, need to guarantee the ideal conditions for this encounter to take place and later for the egg to find a nest.

We have already mentioned that the pituitary gland sends messengers to the ovaries and these respond to the stimulating message by secreting hormones (estrogens and progesterone) that effect changes in the uterus, preparing it for the gestational process.

The production of sperm inside the testicles responds to a similar feedback loop.

Secondary sexual characteristics appear during puberty due to the pituitary secretion of certain substances that trigger the production of *estrogen* in the ovaries or *androsterone* in the testes. Women and men produce both male and female hormones, only they do so in different amounts. It is interesting to note that after the hormonal production decreases, causing menopausal and climacteric changes, there is a certain feminization of the man (distribution of body fat changes, for example) and certain masculinizing of the woman (hirsutism).

Our life, as biological individuals, starts at the moment we are conceived, when male and female gametes (sexual cells) come together. From that moment on, the individual just formed goes through several maturation stages in what we call the life cycle. Each stage is characterized by evolutionary biological, social and psychological tasks, similar to those described when we talked about subtle bodies. Each maturation stage is sustained by the former.

Gametes have peculiarities that make them different from any other cell in the body. Their nuclei, instead of the normal 46 chromosomes, have only 23. Chromosomes contain the genetic material that determine the color of our skin, the color of our eyes, the tendency to be tall or small, to gain weight or be thin; in other words, the features that we inherit from our parents.

110

However, except for the ones determining our phenotype, a gene doesn't express itself without certain given conditions. There are dominant genes and recessive genes, genes that inhibit the expression of other genes and genes that facilitate the same. Most genes linked to a specific disease won't express themselves unless certain environmental conditions exist.

The complete collection of genetic material in the cell is known as *genoma*. One human cell has 23 pairs of chromosomes in the nuclei and one chromosome in the mitochondria, which is called *mitochondrial DNA* (this is not transmitted by the spermatozoid).

The genome project, mentioned above, mapped the chromosomes, identifying the site (locus) of more than 2,000 genes linked to congenital and hereditary diseases.

Genes determine our individual differences and explain, in part, why in the face of similar environmental conditions, our responses differ. There are still authors who have a deterministic vision of the genes who affirm that our heritage has no other choice but to express itself, even though this expression seems to happen at random.

But science has demonstrated that we indeed have some control over hereditary features (maybe not the color of our eyes, but yes, we do, over our height or weight). Our lifestyle inhibits or facilitates the expression of genes. And knowing this, we have even more arguments to take responsibility for our bodies.

Genes do not impose but they express certain characteristics. *Epigenetics* studies the changes in genetic expression that are not linked to alterations in DNA sequences. Genetic expression refers to the fact that even though we may inherit certain features from our parents, specific environmental conditions are required in order for the genes to be switched on or off before an illness manifests in the body.

An example: The only child in his family with a very early awareness of the connection diet-diabetes, Frank, was spared the condition suffered by his parents and five siblings. At age 60 he continues to be very careful of what he eats and has never developed high blood sugar. Although *epigenetics* doesn't completely support Lamarck's concepts, it raises the possibility that *epimutations* as these gene-turning on or off of genes are called, could play a role in evolution.

Genetics, the science that studies how our chromosomes store information, demonstrate that DNA is in charge of telling the cells what to do, when to do it and how to do it, to keep the body functioning. The

DNA somehow *knows* how to respond to the inner environment of the body.

IX. Nervous System: evaluation, relation, response, regulation and connection

It appears that the nervous system was the first one to be considered a communications system. Several of the mechanisms by which nerve cells transmit information and the nature of the molecules in charge of the transmission have been well known for more than a century.

This system is in charge of the evaluation of the inner and outer conditions in which the body has to perform and also in charge of issuing commands that facilitate the relationship with the environment or the maintenance of stable inner conditions. These functions are mediated also by messengers – chemical substances that are known as neurotransmitters or *neuropeptides*.

Remember how we compared the communication between the body and the nervous systems with a *GPS*. We grasp the world through our senses; the brain collects data about the external world using visual, olfactory, gustative, auditory and tactile representations, which provide us with information necessary to respond to stimuli. The inner world – our anatomy – possesses certain sensors (propioceptors) that generate information about our position in space, the tension in our muscles, the blood pressure in our arteries, distension of the stomach, etc,

An axon, a long projection that emerges from the body of a nerve cell (neuron), transmits messages – like a feeling of pain coming from the skin to the brain or a command to move coming from the brain to a muscle. Dendrites, shorter projections that branch out from the neuron body, transmit information to the cell through electrical signals emitted by other neurons through their axons.

A lineal frame of mind, which is limited in understanding of the functions of the body, postulates that information is transmitted in a *receptor – dendrite – neuron – axon* sequence, through special structures called synapses. These are like bridges between axons and dendrites, in which certain chemical substances – *neurotransmitters* –are secreted in the presence of a stimulus. However, in the eighties several researchers found that only two percent of the communication between neurons happens at the synapse level.

112

Neuroscientist Francis Schmitt from the *Massachusetts Institute of Technology*[38] demonstrated the presence of messengers, which he called *information substances* (transmitters, peptides, hormones, factors and proteins) that travel throughout the intercellular fluids looking for specific receptors to deliver their messages. His discoveries imply that communication mediated by the nervous system is even more complex than previously thought. It is similar to the communication seen in the endocrine system. The neurotransmitters are found not only in the synaptic ends but also in the target organs over which the nervous systems acts. In addition, the communication is multidirectional and not bidirectional as was previously thought.

There are billions of cells working within our brain, grouped in teams performing their different functions. When the neurotransmitters are released, they either cross the space in between the neurons or travel to deliver information to different organs in the body. In the synapses, after one stimulus is transmitted, the remaining messengers are removed and reabsorbed by the synapses that initiated the transmission of information, to be later reutilized (this is called re-uptake).

Neurotransmitters affect other systems.

One of the main transmitters is *serotonin*, which is one of the substances in the body that modulates mood. There are receptors for serotonin in different organs of the body (actually, 95 percent of the body's serotonin is housed in the gut). This molecule has different functions: it influences the formation of blood clots and produces constriction of blood vessels. It also plays an important role in sleep, appetite, memory, sexual behavior, breathing, aggression, peristalsis and the function of our glands. However, its most important function is the regulation of perception that gives us a sense of reality.

Serotonin and its role in depression will illustrate how the nervous system works from a dynamic perspective.

While writing an article about depression for my column *Body, Mind & Spirit* in the *Naples Sun Times*, I realized that I have mostly dealt with the topic from a medical perspective. As a doctor, I believed that it was a condition that could sometimes be genetically explained, and often required medication. It is true that at the chemical level, depression is related to the production of brain mood modulators such as serotonin.

---

[38] Mentioned by Candace Pert in her book *Molecules of Emotion*

Later, as a psychotherapist, I realized that there could be certain family dynamics that prompted depression in one of its members, and I started to explore the significance of relationships and unresolved mourning processes as a major cause of depression.

When I became a reiki practitioner and moved towards alternative practices, I realized depression was related to the search for fulfillment and to the loss of the connection to the whole to which we all belong.

Today, I see many plausible explanations for what Freud called *melancholy*, but overall, I see it as a sign of lost balance.

In my current multidimensional vision, I believe that depression manifests simultaneously on all levels of our existence (physical, emotional, mental and spiritual). That's why doctors see it as a chemical disturbance while psychologists approach it often as a reactive mental process, and spiritual guides might consider it necessary to seek God for comfort.

Doctors offer pharmaceutical solutions that include many synthetic antidepressants which inhibit the re-utilization of serotonin at the synapse level. This improves mood, but at the same time, causes a number of unwanted side effects. For example, the medications stimulate the production of insulin, which lowers blood sugar. The excess amount of free serotonin caused by the medication interferes with digestive function and the blood vessels and can affect the heart's valves[39].

Experiments with *LSD* demonstrated that serotonin is involved in the disturbance of perception caused by the drug. Perceptional distortion caused by the methamphetamine known as *Ecstasy* has a similar explanation.

Any disturbance in perception has an influence in the way we think and feel. The different functions of our mind are also interrelated in a way that when one is altered, all the others are simultaneously affected.

An increased secretion of serotonin increases the production of cortisol and adrenaline. This hormones produce euphoria, but the adrenals soon get exhausted and take a long time to recover, causing

---

[39] Some of this information has been taken from the excellent article, *"Serotonin and the Pineal Gland"* by Charly Groenendijk, and is reproduced with his authorization. The article can be found at
http://www.antidepressantsfacts.com/pinealstory.htm. Last time retrieved on February 10, 2008.

health problems. Researchers have found that consumers of Ecstasy might suffer from a type of severe depression that is resistant to treatment because the drug has exhausted the natural mood modulators in the body.

The pineal gland, which is a part of the nervous system because it responds to visual stimuli, and a part of the endocrine system because it secretes hormones, contains the highest concentration of serotonin in the brain. This gland also produces *melatonin*, which is a byproduct from the former, and plays a role in the regulation of the function of all the organs of the endocrine system. While the pituitary gland stimulates secretion of the endocrine glands, the pineal plays as an inhibitor. This is important for stress management. Production of serotonin in the brain is cyclical, changing according to sleep and vigil. Magnetic energy can lower the production of serotonin and melatonin in the pineal gland and this is an important reason to unplug electrical devices in our room while we sleep.

The pineal gland is sensitive to light and it is in darkness that it turns its serotonin reserves into melatonin. This explains why serotonin levels are lower at night.

Some other substances produced by the pineal gland (*triptamines*), seem to be responsible for our most vivid dreams, visions, hallucinations and altered states of consciousness. *Pinoline*, similar to the active ingredient in *ayahuasca*, used by shamans in the Andeans, is also secreted by this gland.

Dopamine and *noradrenalin* are the other neurotransmitters associated with our mood. A dopamine deficiency also affects motor activity, which explains tremor typical of Parkinson's disease and tremor caused by the use of dopamine inhibitors in psychiatry.

When a stressor generates a chemical imbalance, this is reflected in a decreased amount of neurotransmitters. Depending on its intensity, stressors can stimulate the production of stress hormones to such degree that the receptors for these substances become insensitive to new messages. Sleep disorders are frequently the first symptom. Dopamine deficiency is also related to poor endorphin production, which lowers the pain threshold and makes it difficult to experience pleasure. Sleep disorders and languor are typical signs of depression.

When the hypothalamus is stimulated by certain emotional experiences, it produces *Corticotrophin Releasing Factor* (CRF) that triggers the production of *Adrenocorticotrophic Hormone* (ACTH). This messenger makes the adrenals increase production of *steroids*

115

(cortisol) that have a regulatory effect over tissue repair. Increased levels of these steroids are found in the blood of people who have been depressed or anxious for long periods. Their pituitary remains activated.

Someone can get out of their depression with Prozac within three weeks, but statistics show this doesn't mean they won't be depressed again after they quit the medication, especially if the root problem is a social or a spiritual one, related to the illusions that we create about what life should be. Until we examine depression from a new perspective, real solutions don't seem to be at hand. And this brings us to an important question about depression and other ills. What was first the *egg* (chemical imbalance) or the *chicken* (social, emotional, spiritual imbalances)?

The *American Association of Psychiatry* has recognized the anti-depressive effect of walking. Regular physical activity fine-tunes the functioning of endocrine, immune and nervous systems. It seems that the sedative effect produced by the endorphins secreted during exercise restores the lost balance little by little. Different medical magazines, such as *Nature Medicine and JAMA* (Journal of the American Medical Association) have published research results that establish a significant relationship between mood and physical activity.

Yale University experts have found that exercise seems to make certain genes in the brain's *hippocampus* more active, especially one called VGF, a gene linked to a neuronal growth factor. The Yale team supported a hypothesis according to which, in order to effectively and permanently affect depression, changes in the actual structure and associations between brain cells are needed and not just changes in the quantity of neurotransmitters.

There is a fine division of labor within the nervous system. It has two main sections: Central Nervous System (CNS) and Autonomic Nervous System (ANS). The former is in charge of motor and sensory functions.

The ANS regulates vital functions (breathing, heart rate, peristalsis) and makes up part of the *limbic system*, which is in charge of feelings and emotions. It is also a mediator between the brain and the rest of the body in stressful situations. It has two subdivisions: the *sympathetic system* and the *parasympathetic system* with opposite functions in most of the cases.

Body mechanisms like the *fight or flight response* are mediated by the ANS.

116

Neurologist Paul McLean, working at *Yale Medical School* and at the National Institute of Mental Health, was instrumental in developing the theory of the *Triune Brain* (See chart 4).

According to McLean, our brains consist of three different evolutionary layers, each evolving from the previous layer. According to this theory, each brain operates as an independent system, each one with its own special intelligence, subjectivity, memory, and its own sense of time and space. Still, the three brains are interconnected and work in harmony with each other, most of the time.

McLean demonstrated that the neocortex, the most recent evolutionary brain structure (involved in higher functions such as sensory perception, motor commands, conscious thought and language), does not dominate the other layers of the brain. The limbic system, which regulates emotions, can overrule conscious functions under certain circumstances.

There is a similarity between this theory and the Freudian postulation of three levels of consciousness: conscious, pre-conscious and unconscious. Some esoteric traditions have also talked about three different planes of consciousness. In the 19th century, the spiritual master Gurdjieff, for example, said that we human beings had a brain for the spirit, one for the soul and one for the body. Kabala, Platonism and other theories mention three centers: spiritual associated with the head; of the soul, associated with the heart, and physical, associated with the navel or the pelvis.

Author Alberto Villoldo proposes in *Four Winds* that the neo-cortex is at birth like a *tabula rasa*, a clean slate over which we inscribe information throughout our lives. In shamanic practices, Villoldo says, the ritual seems to serve as a formula that transmits encoded information to the neo-cortex. The code eludes the logic mind and thus the information is sent directly to the limbic system that in turn commands the regulatory centers in the *reptilian brain*, to facilitate the healing of the body.

Villoldo's hypothesis supports a multidimensional approach. The information that is programmed in the cortex in the way of beliefs and ideas can manifest in the body as symptoms.

Table 4. McLean's triune brain

Reptilian Brain - Called by McLean the R-Complex	Is the oldest part of the brain and includes the brain stem (medulla, mesencephalon, pons, the oldest basal nuclei - the globus pallidus and the olfactory bulbs) and the cerebellum	Regulates and maintains the body functioning, including growth and tissue regeneration, reproduction and conservation instincts. Commands autonomic functions and muscle balance. Center for habits (this brain is rigid, obsessive, compulsive, ritualistic and resistant to change) Headquarters for aggression, territoriality, rituals and hierarchy. Always active even during sleep. Keeps ancestral memories.
Limbic brain, old mammalian brain.	Includes hypothalamus, hypocampus, amydala and pituitary. It evolved millions of years ago in non primate mammals.	It's programmed with the language of emotions. It drives human experience and is the center where the past is retained and evoked. When stimulated some emotions are elicited: (fear, joy, rage, pleasure and pain) Four basic emotional and instinctual responses: feeding, fleeing, fighting and sexual behavior. Body and gestural language Processes most of the olfactory sensations.
Neo-cortex	100,000 years? Outer layer of the cerebrum, occupies two thirds of the hemispheres. .	Matter is transformed into consciousness. Headquarters for abstract thinking, mathematics, interpretation of sensorial stimuli. Adaptability and maturation. Allows us to respond in a rational way to inherited behavior patterns Spatial thinking. Prevision. Capacity to think about oneself. Verbal and symbolic language (Math, reading, writing) Material expression of emotions in science, music, art. Religion and law (to repress the instinctual reptilian and limbic systems).

Source: Most of the above information comes from Carl Sagan's book Cosmos.

One of my students at the medical school where I used to teach back in my country, had the idea (born from a childhood maternal warnings) that she'd *catch* a strep throat every time she walked in the rain. And she would get sore throats, until one day we discussed the case in light of her newly gained medical knowledge. If bacteria caused tonsillitis, how could she explain that a few drops of rain would prompt it? After she understood she had built a myth, she had no more trouble going out in a little rain. Reason had defeated habit and unconscious programming. Was this programming depressing her immune system and making her more vulnerable to infections?

It seems that along our lives we imprint in our neo-cortex's neuronal circuits a series of instructions, sometimes healing instructions, and sometimes ill-prompting instructions.

Neo-cortex is divided into two hemispheres with distinct functions. The left hemisphere, considered masculine, is the logic brain, the center for verbal communication and analytic thought. It is quite lineal; it organizes facts in a sequence and is fast and impatient.

The right hemisphere, considered feminine, is the symbolic brain. It is the headquarters for spatial and abstract thought and artistic skills. This side of the brain is holistic; it assesses the whole and determines its spatial relationships. It processes what is complex, ambiguous and paradoxical. It is the residence for the creative mind.

We have already mentioned what Thorwald Dethlefsen & Rüdiger Dahlke, Louise Hay and Debbie Shapiro have written about the unconscious symbolic meanings of our illnesses. My hypothesis is that the right brain elaborates the symbols that will manifest in the body as symptoms.

X. Endocrine System: regulation, balance, connection

Many functions in the body are determined by the endocrine system that some people mistakenly take for the reproductive system because it includes ovaries and testicles. The existence of messengers is again evident in the endocrine system. In this case these messengers are known as hormones. Some hormones are also neurotransmitters like *adrenaline*, which is produced both in some nervous synapses and in the adrenal glands.

The capacity of the endocrine system to control from a distance the functioning of remote organs by secreting hormones in the bloodstream has been well known for decades. Its role in the mediation and control of bodily process is also widely known.

The pineal gland and the pituitary, both located in the brain, as well as the thyroid and parathyroid glands, the pancreas, ovaries, adrenals and testicles are part of this system. The stomach, the kidneys and the heart, all have endocrine functions.

I will mention only a couple of the novelties that in recent decades have added to the disquiet of scientists about the endocrine system. First, it now seems evident that the secretion of hormones is not limited to the glands that have been traditionally considered part of the system. Other systems in the body also secrete hormonal substances and there are also single glandular cells that secrete them.

Second, adipose tissue (body fat) functions as an endocrine organ that secretes *adipokines*; among them is *adiponectin* that regulates the metabolism of lipids and glucose. *Adiponectin* also influences the way the body responds to insulin and has anti-inflammatory effects on the cells lining the walls of blood vessels.

XI. Immune system: evaluation, reeducation, defense, regulation and connection

Fifty years ago, we still knew very little about the immune system. In those days, only a handful of illnesses were classified as autoimmune conditions where the immune system doesn't recognize proteins normally present in the body and attacks its own cells. Today, researchers have found that autoimmune responses explain about 10 percent of the diseases that affect the planet's population; among them, diabetes (type I), lupus, multiple sclerosis and rheumatoid arthritis, to name only the most common ones. Ulcerative colitis and even schizophrenia are suspected to have a link to autoimmune responses. Furthermore, conditions such as coronary disease have been related to the efficiency of the immune system in clearing up plaque deposits in the arteries of the heart.

By the end of the 19th century, vaccines were invented; Louis Pasteur discovered germs as the cause of many illnesses, and, later, bodily reactions to specific microorganisms, like the tuberculosis Koch's bacillus, were identified, confirming the existence within the body of the immune system. Immunity was conceptualized initially as a defense army in charge of destroying an enemy, reflecting the predominant martial mentality.

Koch and Pasteur inaugurated an era where all illnesses started to be explained in terms of germs. In the early 1940s, viruses were found capable of generating illness, and the sixties and seventies saw a great boom in virology, when researchers tried to establish a causal relationship between viruses and cancer. This causal relationship has not been confirmed, even though in a few cases there seems to be a strong correlation, as in the case of the *papilloma virus* and cervical cancer in women.

The above-mentioned research efforts were not in vain; science has learned a great deal. Studies have established that human bodies continuously produce cancer cells and that the immune system is capable of recognizing *misbehaving cells*, and of isolating, reeducating and/or destroying them, depending on the circumstances. A clear relationship between cancer and the immune system has thus been established. When the immune system is depressed, cancerous cell growth is out of control.

Since the beginning of the AIDS epidemic, investigators have plunged into studying what exhausts the immune system in these patients, contributing very interesting insights into its multiple functions.

Beyond an army that chases, confronts and destroys invaders, the immune system is presently conceptualized as a self-governing network that participates in the body's learning process, and is responsible for both its molecular identity and the biochemical communication between organs. That's why author Fritjov Capra deems it our second brain.[40]

Different from other bodily systems, which are confined to a precise anatomic location, the immune network penetrates each tissue of the body. It is made of a number of tissues and organs (lymphatic organs), specialized cells (lymphocytes and macrophages or white blood cells) that swim back and forth along the circulatory system during surveillance missions, gathering data to ensure the organism's accurate functioning.

This extraordinary system learns and evolves with experience! From the moment we are born, the immune system learns how to react to unfamiliar agents. It learns to discriminate which molecular features typify bacteria that are usually not present in mammals. It also recognizes the body's idiosyncratic proteins. Vaccines are developed based on the immune system's capacity to memorize how to react to alien proteins.[41]

---

[40] Author Fritjov Capra compares the immune system with a complex inner network almost as important for the self-regulation of the physical body as the nervous system.

[41] A vaccine is administered to expose the immune system to small doses of a virus or bacteria, stimulating the production of certain amount of antibodies, which in the future will be able to recognize and control the same kind of agents in case the body is exposed to them for a second time.

There is also a kind of *natural selection* in the thymus, where only T-cells that have learned to unite harmoniously with other cells in the organism can survive.

The *thymus* is one of the most important organs of the immune system. It is a small gland situated behind the breastbone (sternum) and is fundamental in shaping the way in which the body responds to infections. Half of the white blood cells, which originate in the bone marrow, go directly to the blood stream and interstitial fluids. But the rest of them have to go through the thymus where they become *T-cells*. These have three main roles: to stimulate the production of *antibodies* and other lymphocytes, to stimulate the growth and function of *phagocytes* that ingest and digest viruses and bacteria, and to identify foreign or abnormal proteins.

Many immune system organs function as gatekeepers. This is the case of the lymph nodes (in the neck, armpit and groin), the tonsils and the *Peyer's* patches in the intestine. The lymphatic fluid, or lymph, goes through these *customs* checkpoints where lymphocytes capture particulate matter and microorganisms and decide if they should be granted admission to the system or not. Another lymphatic organ, the *spleen*, is in charge of recycling old and dysfunctional cells.

This amazing system only uses its defensive resources when facing a massive invasion of foreign agents.

Recent research shows that the brain, the endocrine glands and the immune system cooperate and share functions. Moreover, the borders that science had delineated between these systems start to blur, bringing opportunities for new understandings of the body's functioning. Candace Pert insisted in the use of the informatics' term *net* to describe these systems, because it encompasses incessant exchange, processing and storage of information. Most substances in charge of transmitting the information are *peptides*, and recent research has shown they are multifunctional; they accomplish different functions for different systems.

For example, the brain produces neuropeptides that are antibacterial precursors; the immune system has perceptual functions, and the endocrine system produces substances that work as neurotransmitters. Initially deemed as exclusive to the nervous system, the neurotransmitters have also been found in the bone marrow, where the immune system cells are produced.

The three systems are thus, multifunctional. They form a network that exchanges, stores and passes on information, using peptide molecules as messengers. But, also, our physiology is modulated by emotions. Popular wisdom, which results from observations transmitted from generation to generation, has always correlated emotional stress with vulnerability to illness, and science has proven that our thoughts, mood and emotions influence the functioning of the nervous, endocrine and immune systems.

In a nutshell, science is telling us that we can regulate the production and efficiency of our inner messengers (peptides) by adopting healthy lifestyles. It's telling us to eat healthy, have fun and step up.

And therefore, organs communicate

Science has not yet gained a full understanding on how the development, regulation and function of the different endocrine and immune systems' components structure an integrated whole. And even though there is plenty of scientific evidence of the multilateral communication taking place among the different systems of the body, there remain a great number of researchers and health workers who continue to support the idea of a unidirectional, or at most bidirectional, flow of information. For example, doctors continue to talk about a *hypothalamic-pituitary-adrenal axis* to explain the feedback interaction between glands.

But the connections are really intricate. We human beings perceive the environmental challenges through our five senses. Information travels to the *thalamus* in the brain, which distributes it to the appropriate processing areas. If a noxious stimulus has emotional content, the information ends up in an area of the brain's limbic system known as the *amygdala* (which seems to record unconscious memories). The *amygdala* is connected with the hypothalamus, where the emotions are processed. The hypothalamus has been recognized as an endocrine *command central* that sends signals to the pituitary and this gland produces substances that stimulate the production of adrenaline and cortisol in the adrenals. The body responds to the secretion of these substances by elevating our blood pressure and heart rate as the body prepares to run away from or face a threat. This is the way emotions become biochemistry in the body.

Modern physiology books regroup the body systems in a way that reflects more accurately the intricacy of the relationship between systems. Let's talk a bit more about how the nervous system, the endocrine glands and the immune system cooperate and share functions.

Researchers from the Department of Physiology and Biophysics of the University of Alabama in Birmingham studied the capacity of the nervous system to recognize millions of antigens (foreign substances entering the body) to which the immune system can respond by producing antibodies. They identified more than 20 *neuro-endocrine peptides* or their corresponding *ribonucleic acids* (RNA) in the cells of the immune system.

In 1919, Dr. Tohru Ishigami, a Japanese doctor who worked for ten years with patients who suffered tuberculosis, found that the activity of the white blood cells decreased when the patients were experiencing emotional distress. He suggested a connection between immune and nervous system. However, the first evidence of such a connection came when he could demonstrate that the cortisol circulating in the blood produced a depressing effect on laboratory rats (it hampered the production of antibodies). Later on, it was proved that peptide-mediated communication between the nervous and the immune system existed.

Other researchers, like Doctor Elena Korneva from the *Experimental Medicine Institute* in Leningrad, have demonstrated that injuries to certain areas in the *hypothalamus* suppressed different types of immune response. Psychiatrist George Solomon also discovered that damages in the hypothalamus inhibited thymus function.

French researcher Gerard Renoux from the *Tours Medical School* described patients with severe injuries of the cerebral cortex, whose immune systems exhibited decreased activity.

In the seventies, Robert Ader, director of the *Psychoneuroimmunology Research Center* at the Department of Psychiatry, Rochester University demonstrated, with the assistance of immunologist Nicolas Cohen, that the immune system of lab rats could be conditioned. Their experiments caused great skepticism among the scientific community until the results were repeatedly replicated with the same results. Ader is deemed the father of the field of *Psychoneuroimmunology*[42] (PNI).

J. Edwin Blalock, who in 1982 discovered that the immune system produced endorphins up to that point considered to be exclusively produced by the brain, has also been recognized as a PNI pioneer. His studies at the *Department of Physiology and Biophysics*, University of

---

[42] Initially called psychoneuroendocrinoinmunology.

Alabama at Birmingham, helped prove that the three systems, nervous, immune and endocrine, use natural body chemicals to communicate among themselves.

Blalock and colleagues indicated that when leucocytes secrete a substance knows as interferon, which defends us against viruses, they also produced ACTH (*Adreno-Corticotropic Hormone*, which stimulates the production of cortisol in the adrenals) with very similar characteristics as the hormone of the same name produced by the pituitary. This evidence was conclusive to demonstrate that the transmission of information among systems is multilateral.

There are still many questions to answer, and that is what science is about – questions.

Everything is connected

"The fundamental delusion of humanity is to suppose that I am here and you are out there..." Yasutani Roshi

Hopefully, our journey through the bodily systems proved that, given normal conditions, the body is perfectly equipped to perform without a need for medical assistance. We can trust that the intelligent *inner healer* will guarantee that the body finds proper responses to environmental challenges and ensure the maintenance of a healthy state.

But this balance depends on how we care for our body in all of its dimensions. Communication among organs should be optimal to preserve the functionality of the *inner healer*. With proper maintenance, this healer can orchestrate the necessary measures to grant returning to balance after responding to a stressor.

Insufficient or inadequate adaptive responses – which in many cases produce symptoms – are the result of one or more of many factors:
1. Virulence of the injurious stimulus,
2. Persistence of a noxious stimulus,
3. The confluence in time of several noxious stimuli,
4. Deficiencies in maintenance: malnutrition, lack of exercise, inadequate stress management,
5. Genetic factors
6. Conflictive relationships, depression, negative thoughts or
7. Repetitive traumatic events (physical or emotional).

The adaptation processes that lead to homeostasis involve both the subtle bodies and the dense or physical body. We have affirmed that any phenomenon has simultaneous but peculiar expression in each level or dimension of a human being.

We have also emphasized the importance of the flow of information between organs, systems, bodies and between us and the environment. Obstacles to the information flow; obstructions to messengers' functions; invasion by alien usurpers that misappropriate messengers' functions; or the incapacity to accurately interpret the messages will weaken the *inner healer* and induce symptoms. Ingesting or breathing chemical substances of any sort seriously interfere with the innate intelligence of the body.

The popular belief that links emotional stress with illness has been backed up by science. In my practice as a reiki master and a psychotherapist I have many times seen how physical symptoms disappear as the person gained awareness and resolved their conflicts.

I have also seen that people, who take inventory of their habits to then focus on improving the areas where they have not yet adopted the healthiest behaviors, easily find resolution of their symptoms; in many cases without using any kind of medication. It doesn't matter if the symptoms were physical, mental or emotional. When they achieved balance, it manifested in all dimensions.

Multidimensionality has that feature. The principle is that all is interconnected. It doesn't really matter from which perspective one approaches the solution, because the changes in one dimension are simultaneously felt in the other spheres. That's why genetics, microbiology, surgery, psychotherapy and spirituality gurus are all right and achieve results. But especially right are the holistic practitioners who approach the human being in their diverse totality and with a clear awareness that it is essential to do no harm.

The *inner healer* responds to challenges imposed by the environment

At the beginning of the 20th century, Hans Selye at *McGill University* in Montreal described the body's response to stress. He discovered that lab animals responded to the injection of different hormones and toxic substances pretty much in the same way: the adrenals showed signs of exhaustion, the lymphatic organs atrophied and the stomach produced ulcers. To this set of symptoms, Selye gave the name of *General Adaptation Syndrome.*

Selye also observed that this response to stress followed certain stages. The immediate response to stress, the *alarm stage*, prepared the body for physical activity. The body concentrated all of its resources to respond to a perceived threat to the system. Put yourself in the situation of an armed robbery. During the initial reaction, the pupils dilate (to improve sight), our heart beat accelerates, and our blood pressure rises (to improve blood supply that will provide enough oxygen to the muscles as carburant and grant enough energy to work properly and to the brain for mental clarity).

Under the command of the hypothalamus, the adrenals will secrete an extra amount of *adrenalin* and *glucocorticoids* (cortisol) and the sympathetic system will be activated. Each one of these phenomena sets us ready to fight or run away.

Once the threat is over, the alarm stage ends. Now comes the stage that Selye called *resistance* or adaptation. Nervous, endocrine and immune systems return to normal. This is a stage of high performance.

But if the threat persists, the body gets exhausted and loses the capacity to adapt, the production of cortisol continues to be high, the immune system is depressed and the body loses resistance to face the stressor. Now illness can kick in and survival is in danger. This is what happens to hostages who have been kept captive for a long time.

When our life is full of stressors, our body tends to be always in *sympathetic mode*. When the sympathetic branch of the autonomic nervous system is predominant, the body accumulates cortisol and adrenaline that become toxic to the body and our tissues pay the toll.

On the other side, during relaxation, the parasympathetic branch of the autonomic nervous system takes over and the processes of tissue regeneration and repair take place. This also happens while we are doing things that are a sources of pleasure: recreating ourselves, hanging out with friends, eating, resting, meditating, receiving reiki or exercising. And this, once more, demonstrates the intelligence of the body.

Not all of us respond in the same way to the same stressor. Our response depends on hereditary factors, compensatory factors such as nutrition, physical activity and meditation, and also on our capacity to develop resiliency. But the response depends on the way we perceive a stressor, as well. The difference in perception explains in great part the difference in individual responses to the same kind of stressor. And our perception is determined by many factors such as genetic factors, previous experiences, beliefs and support systems.

Besides her work with the molecules of emotion, Candace Pert also became interested in the benefits of meditation to counteract the effects of stress over the immune system. She found that stress hinders the free flow of informational molecules, causing a collapse of the autonomic functions (breathing, circulation, digestion), which hinders the repair and regeneration processes. When stress is prolonged, the immune system fails in its surveillance operations.

Meditation and similar practices not only calm our minds, attenuate our emotions and connect us with the transcendent, but at the physical level, these practices also stimulate the reconnection between organs, tissues and cells, which is necessary for balance and to prevent illness.

The relationship between the functioning of the immune system and vulnerability to infections, allergies, and autoimmune diseases (lupus, scleroderma, rheumatoid arthritis) is not a new concept. But only in recent times, and as the comprehension of the immune system broadens, we have understood that its numerous functions go beyond simply defending the body from attackers. We have also learned that behind a number of conditions such as diabetes, cancer and coronary disease, there is an immune dysfunction.

Inflammation and reparation

The main response of the body to an injurious stimulus is inflammation. Although we often see inflammation as an undesirable event, it actually is a protective response from the body.

Presently, inflammation is usually defined as a reaction of the immune system because the white blood cells are involved in the response. Notwithstanding, from a holistic point of view, all systems are involved in inflammation and, on the other hand, the immune system gets involved in all of the activities of the body.

Without the inflammatory response, the body could not start to repair damaged tissues, face an invader or adapt to inner or external environmental changes. The cascade of processes generated by inflammation guarantees the destruction, dilution and sequestering of a noxious agent, be it a bacteria or a thorn in the skin. Inflammation also serves to alert us about, for example, the excessive use of a joint or a muscular group. Inflammation should be understood as the beginning of the adaptation process to any kind of stressor.

It is essential to understand clearly that the purpose of inflammation (and pain) is to protect our body from further damage, to prevent the spreading of an infection, and to facilitate tissue repair. From this perspective, it is clear that we don't have to combat inflammation but rather to support it with healthy habits that will grant the optimal function of the inflammation modulators. All of the physical healing processes start with inflammation in which all of the systems of the body participate in lesser or greater scale.

When an incident happens in a little village, like a quarrel between spouses, a fight in the bar, or an accident, everybody wants to take a peek. Some will just come observe and assess the situation, others are in charge of informing neighbors of what happened, others are ready to help and a few others maybe will get directly involved in the incident. Then after a while, the law enforcers come in and the order is reestablished. Something of the same sort happens in the body where not all of the elements drawn to the *incident* play the same role but all of them have some degree of involvement.

An expanded or exaggerated inflammatory response is always counterproductive. As we have already seen, the body processes are mediated by *coordinators* that indicate to each of the elements involved that their goal has been achieved and they can withdraw from the scene. If the *coordinator* fails to perform at its best, inflammation might become chronic; it can lead to the destruction of tissues or to exaggerated scarring and even to what is known as autoimmune disease. When the immune system overreacts we might see allergies.

On the other hand, when the inflammatory response is not sufficient, the body might remain at the mercy of misbehaving cells, may develop cancer or might be incapable of responding to infectious agents, and then the person will suffer an infection.

The mediator molecules in inflammation are called *cytokines*. Blood levels of these molecules help to determine the presence of unnoticed inflammatory processes. In general, cytokine levels are higher at night and early in the morning when the levels of cortisol in the body are at their minimum. Cortisol counteracts the inflammatory response, which explains why, for example, arthritis pain can be exacerbated at night (or why doctors prescribe cortisone for inflammation). It also explains why people under chronic stress, with high levels of cortisol produced by the adrenals, or people taking cortisone for an extended period, have decreased immune responses.

What we eat has a direct relationship with the degree of inflammatory response. In the last few decades, several authors have classified food as *pro-inflammatory* and *anti-inflammatory*. Within the first group we have refined foods and all others that tend to increase blood acidity, like dairies and meats. In many cultures people recommend that those who suffer from arthritis suppress these foods from the diet, but only recently science has corroborated the relationship between arthritic flares and certain nutritional habits.

In their 2004's February edition, the University of Tufts *Health & Nutrition Letter* published an article entitled *Anti–Inflammatory Eating* where three medical conditions – high blood pressure, cardiovascular disease and arthritis – are correlated with inflammation and nutritional habits. The article points to the fact that the fats and oils in our diet have a direct influence on the inflammation response because they are precursors of *prostaglandins*. These substances behave pretty much like any hormone and they are also messengers that work on neighboring tissues.

Sixteen types of *prostaglandins* have been identified with functions as different as regulating blood pressure, peristalsis and metabolism. Some *prostaglandins* inhibit the inflammatory response (those that derive from Omega-3 fatty acids found in fish, flax seed, certain algae and olive oil) and others stimulate the inflammatory response (those from animal fat, corn, sunflower and cotton oils). The article recommends increasing consumption of fruits, grains, seeds and vegetables, while at the same time avoiding processed foods (often rich in omega-6 fatty acids) to help keep the inflammatory response in balance.

Fruits like papaya and pineapple are also considered anti-inflammatory because of their high content in certain enzymes (*papaine* and *bromeline*, respectively). The body can utilize them because these enzymes are resistant to the stomach acid. Ginger and cherries are also considered beneficial to help offset exaggerated inflammatory responses.

From the physiological point of view, inflammation follows the sequence that you see in table 5, which explains the manifestations and cardinal signs of inflammation: redness, heat, swelling, pain and limitation of function.

Table 5. Stages of the inflammatory process

1. *Vasoconstriction. By a reflex mechanism, the blood vessels narrow either to stop the bleeding or to restrict the entrance to a possible invader. If we scratch our skin, we will first see a white line that is signaling that the vasoconstriction has occurred. It lasts only for a few seconds.*
2. *Vasodilation: The local tissues secrete histamine that stimulates the dilation of the blood vessels.*
3. *Vasodilation: Now the body requires an increase blood flow to the area and this additional supply explains the appearance of redness and heat. The increased flow favors the arrival of white blood cells that gobble up (phagocyte) detritus and invaders*
4. *Vasodilation causes the small blood vessels to enlarge their pores though which some fluids escape and this explains the swelling (edema). These fluids supply certain substances that contribute to the reparation of the tissues.*
5. *The excess fluid in an area produces certain pressure over the local nerve endings, which explains pain but after a while causes numbness.*
6. *The swelling also explains the limitation of function, which is how the body guarantees that there will be no further damage.*

The lymphatic system, which is considered part of the immune system participates in the inflammatory process. It reabsorbs the fluid that escaped from the capillaries and sends it back to the lymph nodes, where there is a customs checkpoint. There, the *macrophages* (huge white blood cells) gobble up detritus and remains from the invading aliens. While the lymph nodes are working they enlarge and feel sore.

Pain and pain killers

A couple of years ago, I woke up with an excruciating pain in my lower back. Although it was a very limiting pain, I was sure that as soon as I started moving around it would disappear, although it is not clear to me why I would be tense after a good night's sleep. Maybe it was due to a poor sleep posture, I told myself. My denial was based on the scientific fact that movement dissipates tension. However, the day turned into night with no relief.

In the past 15 years I haven't taken medication of any kind, except for a few vitamins and supplements, whenever I've felt my body needed them. So, I ruled out using analgesics from the beginning.

The following day not only the pain was still there, it had worsened and it irradiated to the back of my right thigh. I could hardly bend down to put my socks on and when I tried to go for a walk each step felt like a hammer hitting my lower back. My pelvis seemed to remember contractions during labor.

My medical mind told me I had a herniated disk in my lumbar spine. But what would I do with the diagnosis?

Breathing deeply only made the pain increase, sneezing was agonizing. I knew that a doctor would order X-rays or an MRI, which would confirm the diagnosis and localize the lesion with accuracy. After the tests, the doctor would prescribe muscle relaxers and a pain killer (but I would rather use reiki for both), and recommend a hard bed (which I already had). Taking any kind of medication was out of the question. I had long ago decided not to ingest any chemicals that could disturb the communication between my organs.

I played my relaxation cassettes, administered reiki to myself and made an online healing request through the Distant Healing Network (the-dhn.com).

I could hardly sit in a chair, so I straddled my kneeling chair, the one I use on my computer workstation most of the day, making frequent pauses to give rest to the muscles in charge of that posture. I also modified my nutrition to add extra anti-inflammatory foods like tart cherries and natural analgesics like green peppers. I used vitamin C and Complex B to reinforce my connective tissue and support the *inner healer* in the process of repairing any injured nervous tissue.

I took magnesium pills to keep cartilages flexible and bought ripe pineapples that are rich in the anti-inflammatory enzyme *bromeline*.

Although the pain and the muscle spasm were limiting, I knew they served the purpose of protecting my tissues from further damage. That's why I decided to respect the pain. I knew that Reiki and relaxation techniques would keep it within a bearable range and pain would guide my recovery. I knew the pain was a body alert that I needed to listen to.

A few months before, my cousin, after having a herniated disk, was prescribed with a strong dose of cortisone and analgesics. His pain was gone in hours and after a few days he felt so well that he decided to play soccer with his vigorous 8-year-old son. Because the body didn't have enough time to heal, he caused himself an even more serious lesion that rendered him in a wheelchair for a while.

In my case, after a few days, the pain was almost completely gone but the spasm continued to protect the spine for several weeks. I avoided driving, running and bearing weight.

Rolling on the floor, Tai-Chi-like exercises, soft stretching and moving mindfully helped my body take care of the problem while keeping my muscles flexible and relaxed.

In six months I was completely back to normal, but still now, the body continues to remind me with certain soreness in the same area that I should not bear more weight than necessary, that I should change posture frequently and that pauses are desirable.

If the purpose of pain is to protect the body from further damage, we need to listen. Usually, only when the pain is acute and limiting we are forced to pay attention. In most cases, we just ignore the pain as when we have been sitting for hours in front of a computer and the overloaded muscles are asking for a change in posture.

Because we ignore pain, our muscles get tense and develop spasms, which contribute to deform our posture causing misalignments. If we stop using groups of muscles they will shorten and weaken and in time, unattended areas become vulnerable to injuries.

Most of us run looking for medication to alleviate a symptom. But is it really necessary or desirable to counteract a normal defensive bodily response?

I think it is preferable to listen to our body. Forcefully silencing the body cuts off our relationship with it.

If necessary, natural therapies such as reiki, Trager, relaxation techniques and acupuncture will help reduce pain without the undesirable secondary effects of medication. Only in extreme cases should we go for pain killers, and always under a physician's advice.

When we suffer a lesion, the body alerts us by producing pain. There are receptors in our body that detect painful stimuli. These, called *nociceptors,* are nervous fibers that communicate with the brain when stimulated. The message starts at the *nociceptor* level. The membrane of the neighboring cells becomes more excitable to the point of producing electrical discharges even without the need of stimuli. The result is an increase in sensitivity, both to pain and innocent stimuli.

The associated limitation to pain caused by trauma or an injury has the purpose of preventing further damage. For example, in the case of a bone fracture, pain advises us to avoid leaning on the injured part, preventing tearing of the nearby tissues of the splintered bone.

In severe cases like a fracture or a herniated disk, pain is so acute and limiting that we have no choice but to listen. In other cases, we are so captivated with the task at hand that we ignore the pain we are experiencing.

If we listen, then we will hear how pain talks to us. It will tell us about the tension accumulated as a result of staying in the same posture for a prolonged time, contracting the same group of muscles to fight gravity or about the building up of irritation caused by a repetitive motion.

Muscles that have been working to keep our posture expect to be relieved by an alternative group of muscles and announce their desire to shift with discomfort or pain. If we listen to the body, all we have to do is to change posture and the tension will be released. If we don't listen, the muscle resents the negligence, gets inflamed and the pain becomes chronic. The muscle tension becomes a spasm, it deforms posture in a permanent way and with time our bones get out of whack, some muscles stop working and might even shorten. One day we find out that certain areas of the body have weakened so much that we have become vulnerable to injuries. This is typical of modern life, which is extremely sedentary.

Pain is one of many feedback responses that keep the balance of the body. The messenger molecules that transmit pain are *prostaglandins, bradikynins* and *substance P.*

Histamine and *serotonin* also participate in the process. Up until recently, scientists thought that transmission of information was through synapses. But now we know that only a small percentage of the message is transmitted via nerves and synapses, through the spinal cord up to the thalamus and the cortex in the brain, where pain becomes a conscious message.

A substance called *Capsaicin*, found in green chili peppers, stimulates the nerves in such a way that it counteracts *substance P* and alleviates pain in a natural way.

The body also produces *endorphins* to inhibit pain. Many people enjoy a deep massage even when it causes pain because it stimulates the secretion of endorphins that eases the pain and feels good. But the relief extends only for a few hours, after which another massage is

134

needed, creating a "catch-22" situation. We may have developed an addiction to the opiates produced by the body in response to pain and relaxation induced by massage, but we have not resolved the issue that brought us to therapy in the first place.

When the pain is of emotional origin, the endorphins also play a relieving role. If we loose a dear one, the endorphins will help us stay in the initial state of denial that protects us while we can start processing and accepting the loss.

Many people deem it necessary to get pharmacological support to alleviate any symptom or an illness. If we think it over, we will ponder when it is really necessary to counteract a natural defensive response from the body. Using pain as a guide, we could dialogue with the body and find the boundaries within which we can move without further injury.

When we make pain disappear, no matter at what cost, we are ignoring the body, forcefully silencing it, cutting off our relationship with it.

There are many natural ways to mitigate pain without interfering with its function. One of them is reiki. I've heard many testimonials from practitioners and patients who have experienced relief for a headache or a stomach pain after a reiki session, even if the treatment lasted only for a few minutes. In my experience, reiki is faster than aspirin. On some occasions the pain is completely gone and at other times, the pain is just mitigated until its purpose is over.

Among other alternative methods that have proven effective at treating pain are acupuncture, craniosacral therapy and hypnosis.

If pain is the result of illnesses that require medical treatment (like appendicitis or a brain tumor) it will not be masked by reiki. In terminal patients, reiki sedates and alleviates pain, but of course in these cases, pain killers are also necessary and need to be prescribed by a physician.

When we neglect to listen to or when we abuse the body, it responds by exaggerating the alarm it rang, which explains why some people experience symptoms again after they stop medication.

Suzanne E. Simmons from the *National Headache Foundation* (www.headaches.org) advises affiliates against abusing analgesics. "[*Analgesics*] may decrease the intensity of the pain for a few hours; however, they appear to feed into the pain system in such a way that chronic headaches may result. The medication overuse headache (MOH) may feel like a dull, tension-type headache or may be a more severe migraine-like headache. Other medication taken to prevent or treat the headaches may not be effective while analgesics are being overused. MOH can occur with most analgesics but are more likely with products containing caffeine or butalbital."

The pain medication initially increases the levels of serotonin but its continuous use leads to a decrease of the neurotransmitter and more pain, generating what is known as rebound migraines. The cycle can be only broken when the medication is stopped.

Health and illness

What we have said so far should suffice to substantiate a new way to understand health and illness. Absence of illness is not good health, the same way that we wouldn't say that we have proof of the resolution of a marital conflict when the spouses no longer talk to each other. There are moments when there are not many frictions in our life and we produce no symptoms. This doesn't mean that our relationship with our body or our environment is the best.

Within any living being and in nature in general terms, there is a constant struggle between constructive and destructive forces. According to the second law of thermodynamics (Law of increased entropy),[43] the world is inherently active and whenever the distribution of energy is out of equilibrium, there is a thermodynamic force that spontaneously tries to correct imbalance[44].

At the cellular level, cell injury and death are compensated for with regeneration and reparation of tissues and at the systemic level the body strives to maintain homeostasis. At the energy level there is also an ebb and flow, manifested in blockages that the body can overcome. Notwithstanding, we all age and occasionally get ill while our culture tells us that we should fight aging and illness. At most, we can delay

---

[43] While quantity remains the same (First Law), the quality of matter/energy deteriorates gradually over time. Energy is used for productivity, growth and repair and in the process, usable energy is converted into unusable energy. Thus, usable energy is irretrievably lost in the form of unusable energy. "Entropy" is defined as a "measure of unusable energy within a closed or isolated system" (the universe for example). As usable energy decreases and unusable energy increases, "entropy" increases. Entropy is also a gauge of randomness or chaos within a closed system. As usable energy is irretrievably lost, disorganization, randomness and chaos increase.
Source: www.allaboutscience.org/second-law-of-thermodynamics.htm. Retrieved on February 12, 2008
[44] To know more about entropy, visit: www.entropylaw.com/entropy2ndlaw.html

the aging process with healthy lifestyles and grant that our elder years will be comfortable.

At least 75 percent of illnesses are self-limiting, which means that they run a definite limited course and they actually do not require any kind of intervention. The body can efficiently battle an aggressor, heal a wound or restore lost balance. We can easily see this with the most common childhood rashes like chicken pox or measles. Another example is the common cold that has no other treatment than symptomatic. And according to the most recent reports, even this symptomatic treatment should be limited to what grandmas used to advise: more fluids, lots of rest, saline solution for stuffy noses and loving care (including chicken soup!)

In 2007, U.S. health experts urged the federal *Food and Drug Administration* (FDA) to consider banning the sale of over-the-counter multi-symptom cold medicines for young children. During 2004 and 2005, three children had died and an estimated 1,519 children under age 2 were treated in emergency rooms for problems associated with cough and cold medications.

"Because of the risks for toxicity, absence of dosing recommendations, and limited published evidence of effectiveness of these medications in children aged under 2 years, parents and other caregivers should not administer cough and cold medications to children in this age group without first consulting a healthcare provider and should follow the provider's instructions precisely," a Center for Disease Control report said.

Overdoses were responsible for some – but not all – of the adverse events. All three of the dead infants had high levels of pseudoephedrine, a commonly-used nasal decongestant, in their blood, according to CDC.

Analysts, who study statistics and concern themselves with social costs of illness, seem to agree that in the United States people utilize medical services more often than necessary and that doctors prescribe medications that are not essential. One of my friends says this is the outcome of a country that has come to be ruled by the fear of malpractice lawsuits.

Another consideration that we need to make is that if stressors are a constant part of life, why is it that the same kinds of stressors make some people ill and not others? What determines individual responses to stress? What is it that defines when, facing the same kind of stressor (say for example, a cold virus) sometimes we get sick and sometimes we don't?

Let's ask again. What determines our response to environmental challenges? What makes a symptom prevail and what designates the system and organ that will be affected? We have already talked about genetic factors, labeling through diagnosis, programming illness in our neo-cortex and the importance of perception.

Several theories explain the onset of illness from non-holistic perspectives. In this book we have focused on explaining how when stress overcomes our resources, it becomes the cause of illness. Many kinds of stressors: physical, biological, electromagnetic, chemical, emotional and mental, are part of our daily life. In health, the body is well equipped to adapt to stress and cope with it. But symptoms are not, of course, the result of a mathematical equation where stressor = illness.

As we have already discussed, from an environmental point of view, poor sanitary conditions, pollutants and virulent microorganisms explain disease. From the body's perspective, our immune response, our mood, genetic predisposition, biochemical changes and nutrition contribute to health.

Because of all the factors involved, I have found it difficult to formulate a definition of health to fit my current understanding of the human body. Even books talking about an *inner healer* (or similar concept) or about vibrational medicine or about illness as an opportunity to achieve personal growth, seem to focus on the efficacy of this or that cure or diet, and the accuracy of this or that device. Most books offer formulas, advice, panaceas and truths but not a definition of health and illness that could help pull us out the old paradigm.

The old World Health Organization definition[45] seems utopic. *Total physical, psychological and social wellbeing*? When would we achieve that? It leaves me with the feeling that the common factor can then be no other than illness.

Without a holistic vision, all possible definitions of health and illness are faulty. Without a dynamic vision that understands the body's struggles and successes adapting to the environment and maintaining balance, all explanations seem partial truths, only expressions and formulations of the same reality but seen from different perspectives

---

[45] "Health is a state of complete physical, mental and social well-being and not merely the absence of disease or infirmity." WHO

that do not encompass the whole picture. A holistic approach would talk about evolution, ecology, adaptation, resources, balance, alarms and defense mechanisms.

In the process of learning and adapting to environmental stressors throughout our life, our organism *programs* emotions, reactions, behaviors and even physical symptoms that help it deal with stress. We could say that we *learn* health and illness, and once learned it becomes a *script* the body mindlessly follows. Our energy – spiritual, emotional, physical – creates the script and our thoughts and feelings become the modulators of our well-being.

When doctors establish a diagnosis, especially if it labels a chronic condition, we start revolving around the illness. We visit doctors once or twice a month, ask ourselves what to eat, what's the proper amount of rest we should have, what kind of supplements to take, what activities to avoid. We read and learn about illness. We talk about it. A diagnosis becomes a complement to our identity, and it may even provide us with what Sigmund Freud called *secondary gains*, which include obtaining people's compassion and understanding and certain protagonist role. At a conscious level we're battling the disease but, unconsciously, other forces are at play. As we said before, a cold might provide a perfect excuse for an overly responsible person to slow down and isolate himself.

I can't agree with those thinking that a call to become responsible for our own health equals blaming the victim. I think an illness might be an opportunity for personal growth and increased awareness if we don't let ourselves become its victims.

The way in which our thoughts, attitudes and perception influence our wellbeing is demonstrated by the *placebo effect,* which has been compared to the power of suggestion. Most people view the placebo effect with certain disdain. In research in many cases, the placebo effect has value = 0 for the researcher.

Placebo-controlled studies deem medication effective if the outcome is greater than the observed using a placebo. Because the researcher's belief in the value of treatment may affect the outcome, the studies are usually "double-blind," where both the patient and the researcher are unaware of who is receiving medication and who is receiving the placebo. But the placebo effect is not 0 and that explains why many studies show that at least a quarter of individuals given placebos experience relief of symptoms and a measurable improvement in their condition. The effects are related to their perception, their thoughts, and their faith.

A study on depression medication (Khan A, Warner HA, and Brown WA, 2000) argues that up to 75 percent of symptom reduction and suicide risk reduction is due to the placebo effect rather than the treatment itself.

Objective? Subjective? In a multidimensional approach to the body there is actually no difference. *The observer influences the observed, the observed influences the observer*. Among the physiological changes that might explain the effects of placebos are an increase in brain blood flow and neural activity associated with faith in the medication the patient is taking, while pain-relief with a placebo is explained by mood-induced release of sedating endorphins in the body.

Norman Cousins authored *Anatomy of an Illness*, based on his battle with a chronic and progressive disease. He said medication is not always necessary to cure the patient, but faith always is. The placebo effect, he said, has allowed medicine to evaluate and understand the relationship between mind and molecular changes in the body.

Renowned authors like Louise Hay, Eckhard Tolle, Jeane Carper and Brough Joy became famous when they published their healing experiences. They passed the test of illness by transforming their lives.

After 40 years of a wellness movement focused not on illness but on health, mind-body practitioners assure that the body can be cured through the mind and vice versa. *Psychobiology, biological medicine* and *psychoneuroimmunology* have demonstrated in tangible ways that mind and body are indivisible. There are innumerable authors that have contributed to that knowledge: Wilhelm Reich and Alexander Lowen (bioenergetics), Ida Rolf (Rolfing), Milton Trager (Psychophysical integration), Moshe Feldenkrais, John Upledger (Craniosacral therapy), Christine Caldwell (Somatics), Franz Pearl (Gestalt therapy) and many more.

Psychobiology, biology medicine and psychoneuroinmmunology deserve special mention because they have demonstrated through scientific research, the indissolubility of the mind-body binomial. Thanks to developments in these fields a new tendency was born, called *integrative medicine*, of which Dr. Andrew Weil is a well-known pioneer. Other medical doctors such as Deepak Chopra, who was trained in the United States but kept connected with his roots through Ayurvedic medicine, have contributed to bring new light to some of the truths that before seemed incompatible with science.

Doctor Larry Dossey proposed that we are entering a *third-era medicine*, which is all about consciousness and connection with the transcendental, where healing ultimately occurs as a result of love, intention, and visualization.

Several studies designed to evaluate why certain medical treatments are more effective, have established that the outcome doesn't depend as much, as it might be thought, on precise diagnoses and accurate medical prescriptions. On the contrary, it has been found that the success of a treatment lies in the disposition of the person to respond in certain ways to illness, his perception of the situation, the support system the person can rely on and even on the quality of the attention received from health providers.

Let's welcome the new understandings of the relationships between health, environment, thoughts, perception, feelings, attitudes, nutritional habits, breathing patterns, spirituality and lifestyle.

I can't insist enough that the focus of medicine cannot continue to be illness but must become the human being and our marvelous *inner healer* requiring support to perform optimally.

In 2002 I visited a chiropractor who, through computerized spine analysis, concluded that there was interference to the transmission of nervous impulses along my spinal cord. Thirty years back I had a serious car accident in which my neck most probably suffered a whiplash injury. Since the accident, I had had the almost constant sensation of bearing weight on my shoulders and my neck was usually quite tense. The chiropractor showed me the X-rays. There was a subluxation of atlas over axis and an anterior displacement of the sixth cervical vertebrae (spondilolistesis).

Just after the first adjustment I had a flashback where I revived the accident. I "saw" Bogota's gloomy morning, the trees on the sides of the avenue and the bus against which I collided. I felt the cold and heard the sound of the metal crashing. And suddenly I was experiencing the fear that for some reason I was unaware of at the time.

Several authors have documented cases of patients reviving traumatic events during manipulative treatments or who experience physical discomfort while recalling traumatic memories. Are some of our memories kept in our body tissues? Some researchers believe so. In her book *Anatomy of the Spirit,* Caroline Myss proposed that each organ and system of the body is calibrated to absorb and process specific psychological emotions and energies.

Researchers who have studied this topic report that an individual can relive traumatic experiences with bodily sensations and reactions similar to the original ones, after which comes a physical relief.

After the emotional release at the chiropractor's office, my neck relaxed and the pain was gone completely for at least two weeks. Subsequent adjustments brought relief but never as dramatic as the first time. Muscles tend to return to the restrictive patterns that for a long time maintained the changes to the structure. I believe that freeing the energy repressed in the neck, more than the adjustment of the subluxation, was what brought about the miraculous reprieve of the chronic spasm during the first two weeks.

For years, the pain in my neck, upper back and right shoulder changed in intensity, increasing when I had stressful situations in my life. Some authors maintain that intense physical or emotional traumatic experiences permanently disrupt the response of the body to stress and thus, in new stressful situations, we will again show symptoms in the areas that have been made vulnerable by trauma.

A couple of years ago, another chiropractor with a holistic approach found that I also suffered a subluxation of the acromio-clavicular joint (between clavicle and shoulder blade), which he adjusted, treating me with vibration, laser and other alternative modalities and I've seen complete resolution of the shoulder pain and limitation while the neck only bothers me when I neglect it.

Osteopaths believe that tissues subjected to trauma become quite vulnerable to any kind of tension because they become a weak point in the system. If our inclination is to consider reasons different from the physical, then we can see a weak point as a symbol for the principal challenge that our evolutionary path is presenting us and it will certainly tone down after we learn what we need to learn.

Biology scholars have demonstrated that after a prolonged or devastating trauma, stress hormones (adrenalin, cortisol) can inhibit or damage the hippocampus[46], compromising the conscious memories of such event. This could explain the dissociation that usually happens after traumatic experiences such as sexual abuse or war events.

Candace Pert's work has also served the purpose of explaining how the nature of all phenomenological experiences is linked to a

---

[46] The hippocampus acts like a gatekeeper of sensory information. To form long-term memories, the hippocampus needs to go through a chemical process that strengthens the synapses. People who suffer Alzheimer's have a malfunctioning hippocampus.

certain state of consciousness. We tend to retrieve information in the same state of consciousness in which we learned it. In 2003, the *Alcohol Research & Health* magazine published the article *What Happened? Alcohol, memory, blackouts, and the brain*, by Aaron White citing Goodwin et al.'s experiment performed in 1969. A group of students learned complex mathematical problems under the influence of alcoholic intoxication and they could only remember these problems when they were again under the influence of alcohol.

Psychobiology has proven that memory, learning and behavior are all influenced by substances known as *neuromodulators* that can store information in the brain.

Peter Levine in his book *Waking the Tiger: Healing Trauma* says that when we human beings are in a stressful situation, we have to face the *fight or flight* dilemma, while the animals act by instinct. This makes us vulnerable to the effects of trauma. If, instead of acting, we freeze, the energy that was not discharged is kept in the body and can, eventually, appear as a symptom or behavior.

After having reviewed these elements that explain the appearance of symptoms, I want to propose that *health **is a state of maximum consciousness and interconnectedness that allows our soul to freely flow with no attachment to what we were in the past or what we possess or what we might become or what we wished for that never happened.*** This special state of consciousness has a correspondence in each dimension of our being; at the biological level as optimal communication between organs, and in the subtle bodies as the unobstructed flow of vital energy and spiritual harmony. In this state, the *inner healer* can perform its role and our minds can accept the ebb and flow of our natural cycles, listening to and learning from the body.

What we call *illness*, on the other hand, would be that other ***state in which an imbalance manifests itself in the most vulnerable dimension or dimensions of the body, with symptoms that have a purpose: to contribute to adapt to the environment and restore the lost equilibrium.***

To maintain balance, the response of the body to stressors is propped up by the three basic pillars we have already mentioned: *nutrition, physical activity and stress management.*

Stress management includes the way we breathe, the way we relate with others and the environment, and the precepts we abide by.

The multidimensionality of the body implies that the changes taking place in one of the dimensions of our body will have a correspondence on the others. An example would be this person who, as a result of working with her emotions, changes the perception of herself, increases her self-esteem and starts taking better care of her body, at the same time that she improves her concern for the planet that she now doesn't want to keep polluting it. Another example is a person who has gained enough discipline to go to the gym every day, which has help him balance his thyroid hormones and overcome depression.

Inquisitive minds formulate new theories

I feel that new generations of health professionals have open and receptive minds even though when they attended school, their curricula tended to be based on the exclusive acceptance of what has been corroborated by experimental science. Within this paradigm, personal experience tends to be dismissed.

But we cannot put aside our important legacy. History has shown us that many theories were deemed ridiculous or implausible in their time because they could not be corroborated, like the roundness of the Earth or that adult human brain didn't have the capacity to grow new brain cells (we do!)

A hundred years ago, Einstein postulated the *Theory of Relativity* according to which there is a fourth dimension, in which time and space form a unity. Einstein deduced something really simple: because it's impossible for different observers to have the same vantage point, each one would draw conclusions according to their own perspective and thus our evaluation of reality is relative. Notwithstanding, instruction in most schools, elementary, secondary, undergraduate and even graduate, is still based on a skewed version of what Newton postulated in the 16th century (his ideas about energy are left out). He had a notion of absolute time, absolute space and absolute motion and believed that nature could be objectively described.

Einstein's genial ideas didn't come to him in a laboratory but while he walked though open country in Italy. He demonstrated that you engender science when you methodically approach phenomena with questions in your mind. These questions include a reappraisal of the laws, theories and concepts that, formulated by others, have already gained consensual acceptance and become mainstream. If we confine ourselves to what has already been measured and experimentally founded, we may ruin the joy of inquiry, restricting our right for unlimited curiosity and condemning us to arrogance and bigotry.

The formulation of new thought systems like *Chaos Theory* to study complex systems has refreshed many scientific fields where existing theories were short in plausible explanations. Even though the word *chaos* evokes the idea of confusion, in mathematics this theory's subject matter is the unpredictable, offering a view of the universe as an interconnected whole. The theory describes the behavior of certain systems (like the weather) that may be highly sensitive to initial conditions. This sensitivity manifests itself as an exponential growth of perturbations from the initial conditions.

Chaos Theory has its origins in Edward Lorenz, a meteorologist who wanted to explain why infinitesimal variations in the input data would have computers render such differing weather forecasts. In Chaos Theory the behavior of certain systems appears to be random.

Another conclusion drawn from this theory is that because of our interconnectedness, we are all co-responsible for the universe. If, in looking at the history of our planet, we asked how our lives would be if we could change some historical events, we would immediately see the concatenation of events and the responsibility that we bear.

If each one of us (Population of the Earth the minute I write this is 6,669,116,240) threw one little paper on the street... just imagine how littered the streets would be! Change one light bulb in your home and contribute to offset global warming. "If every household in the United States replaced one regular light bulb with an energy-saving model, we could reduce global warming pollution by more than 90 billion pounds over the life of the bulbs; the same as taking 6.3 million cars off the road," according to the Environmental Protection Agency.

Reductionist visions that postulated, for instance, that by understanding the germ we would understand the illness, or that mind and body belong to different areas of study, are giving way to a holistic, integrative, perspective. Curiously enough we are going back to many of the things that Hippocrates postulated 2,500 years ago. We have to acknowledge of course the important roles that geniuses like Newton and Descartes played in the development of thought and science, making the emergence of analytical disciplines and the knowledge of the biological causes of illness possible. But the task of our days is to work towards a unifying theory to understand the universe.

New theories, such as Chaos Theory, the morphogenetic field theory, quantum physics, energy healing and the holographic theory of the brain, can be applied to a broader, cosmic perspective. Our commitment is to understand the diverse dimensions of health and illness and to transform our practice, our pedagogic duty and self care to provide the planet with therapeutic solutions.

Throughout history, scientists have contributed to the new concepts on health and illness that we have presented in this book. Among them is Claude Bernard who by mid 19th century discovered, for example, that the liver keeps sugar reserves which are discharged into the bloodstream when the body demands it. His studies led him to understand what he called the body's *milieu interior,* an organic harmony that maintained conditions constant in the presence of external changes, and that is dependent upon the capacity of the body for self-regulation.

Walter Cannon, in his book *Wisdom of the Body,* and in line with Bernard's discoveries, was the first one to postulate the idea of homeostasis or internal equilibrium.

The father of psychoanalysis, Sigmund Freud, contributed the understanding of the existing relationship between body and mind when, while studying cases of hysteria, he not only found *conversion* of psychic traumas into physical symptoms, but also found his patients deriving secondary gains out of their illnesses.

Psychiatrist Franz Alexander shocked the medical establishment when he ascertained that some chronic illness such as arthritis, gastritis and colitis, with no known biological causes at the time, had a close relationship to stressful situations. He is considered the father of psychosomatic medicine.

We have already mentioned Selye, who explained the mechanisms by which stress affects the organs. Lawrence LeShan considers that there is a certain type of personality linked to cancer. George Vaillant described how individuals who use immature coping styles (denial, dramatization, etc) get sick more often than most people.

Gerber, in *Vibrational Medicine,* says that perhaps the key to treating recurrent health problems is not in mending the issue with a physical solution but in correcting the "energy organizing patterns" that direct the cellular expression toward dysfunction.

And we may remember that science has found that the brain produces immune modulators and the immune system has sensory functions.

146

To summarize, medicine has evolved towards the understanding of a dynamic body, with organs that communicate among them, with an intelligent *inner healer*.

The body speaks to us

"What do we get sick for?" This is a question that belongs to the mind-body or *era-two* medicine.

If we say that Mr. Smith suffers from high blood pressure to solve a chronic conflict in his married life, many will look at us in amazement and incredulity. Notwithstanding, science is getting closer every day to this kind of explanation that invites health professionals to practice medicine in a different way.

We have maintained that our body is multidimensional and that when we are out of balance the body will give us signals in each of its dimensions and in its own peculiar way. Emotions have a corresponding expression at the physical or molecular level. An emotional state goes along with certain set of thoughts and our spirit seems to shrink or expand according to the thoughts that we unconsciously nest. But we are so used to draw cause-effect explanations that it seems easier to say that our emotions made us sick or that our thoughts caused certain emotions, when actually what is happening is that they are simultaneous. Emotional and physical symptoms are more evident and that is why we tend to focus on them. However, when something affects us, it affects the whole of us, even if we can only see part of the picture.

We know now that *unhappiness attracts viruses*. We have already explained that stress hampers the functioning of the *inner healer*. But why is it that an individual develops a certain set of symptoms and not others? And why now? There are no conclusive answers, but we can explore a path to them.

So-called psychosomatic medicine pointed to a symbolic relationship between certain illness such as dermatitis (skin disease), arthritis (joint disease) and gastritis (digestive disease) and the repression of aggressive tendencies. These emotions would then become apparent at different levels: from the most superficial ("skin-deep" emotions) to the deepest ("a lump in my stomach").

I met a surgeon who in the zenith of his career started to suffer arthritis. *What did he get sick for?* A few years after his diagnosis and after a tenacious search for solutions, he rejected pharmaceutical drugs that were causing severe side effects and embraced alternative medicine. He mourned the fact that he no longer could continue to be a surgeon but found that *illness had healed his life*! He was happy to leave behind a stressful life that didn't allow him time to be the totality of what he wanted to be, and now he had a more significant life where even his family relationships greatly improved.

A little girl raised by a very controlling grandmother was brought one day to my office. She didn't want to eat. It was interesting that she didn't have problems eating at the neighbor's house. It seemed that she refused to eat at home in order to assume some control and autonomy and counteract the excessive control that others exerted over her body and life at home. She regained her appetite after the family changed some of their relational patterns.

I have seen how an accident suffered by an adolescent can lead to a redefinition of parent-child relationships. A case of a cystic fibrosis patient who developed a psychotic crisis made me think that he had gone *out of his mind* to better deal with his fear of death. He developed a delusion through which he fed the fantasy of being saved by doctors. He also made two suicide attempts that seemed to have served the purpose of freeing his mother from the burden he felt he was for her. And of course, we are talking about unconscious reasons.

While teaching a class about this subject at the University of Cartagena in Colombia, the students challenged me. They couldn't agree with what I was saying. We were sitting in a circle and I asked all of them to tell me the last time that they had been sick. By examining their secondary gain they all could easily see the relationship between unconscious motivations and illness. The last one to speak had suffered a car accident and had fractured his left arm. He was happy to demonstrate to the group that I was wrong. Accidents happen! Then I asked him to tell me how the accident had happened and he told this story:

"I was having an argument with my father because he didn't want to lend me some money I needed. So I left the house angrily and took the car and ran a red light."

Then he realized that the cost of the accident, which his father had to pay for, was very close to the amount he had been asking for.

A coincidence?

I do believe that it would be of great help for the health professional if he asked, even if only to herself, *what for?* Maybe with an understanding of this dynamic, the doctor would be able to predict health risks.

In a systemic, dynamic frame of reference, it is easy to understand that when a family member gets sick, the whole family is in some way affected. But also, on many occasions, a patient's illness serves the purpose of maintaining the balance in the family. In fact when the patient gets well, another member will fall ill if the family dynamics have not changed.

The hypothesis that cancer patients have unresolved business with others and that forgiveness might lead to spontaneous remission of tumors seems interesting. A group of physicians in Germany, following the interesting work of Dr. Ryke Geerd Hamer, has treated cancer patients with psychotherapy claiming between 90 and 97 percent success rate! But are unresolved business the cause of illness? Is illness a solution? Or both?

I suggest the questions in table 6 as a starting point to look at an illness from a new perspective. Some of the questions are inspired by Dethlefsen & Dahlke's book, which I recommend reading.

I know we are not always ready to scrutinize our lives and that, depending on how you take them, you might feel some resistance to these questions. You might feel that you're been accused of forging the illness or of causing it. But there is no faking or fault. The mind has become body to deliver a message that we could clearly understand if we paid attention. It doesn't translate that we cause our own illness, but that our multidimensional body needs to resolve its frictions and regain balance and for that, it has to speak up. It has to tell us that we have probably been postponing the resolution of a conflict at the level at which it originated (mental, spiritual, emotional) and now it becomes physical, visible, palpable, so that we face it and resolve it.By seeking the answers, we may begin to use on our benefit and to stimulate the self-healing capacity of the body.

Table 6. The whys of the symptoms

- *Why do I get the symptoms in this precise moment?*
- *What is in my life in this moment that "makes me sick?"*
- *What are these symptoms telling me about my lifestyle, my nutrition, myself? Is my immune system weak? Am I doing what I must to keep balance in my life?*
- *Are my symptoms the result of self-destructive behavior? Do I love myself?*
- *What is the function of the part of my body that is affected?*
- *What is the correspondence between this function and other levels of my existence? For instance, if you're suffering from a respiratory problem, you make ask yourself: what is in my life today that suffocates me? If your blood sugar is high, ask yourself, is my body compensating for the lack of sweetness in my life?*
- *What are the unconscious secondary gains from this illness?*

Reiki, art of healing

Reiki is a Japanese word where REI means *universal life force energy* and KI (the equivalent of the Chinese word Qi or Chi, Indian word Prana and close to Freud's "libido" or Reich's "orgon") means individual vital energy. When Rei and Ki flow together, we are whole, healthy. Blockages to the flow of energy at any level of our being manifest as disease. Symptoms help us learn about ourselves and become a tool for personal growth.

Reiki is based on the funneling of *Universal Energy* through the practitioner via the recipient, to facilitate somatic, mental and spiritual healing. As said before, healing is more than the popular notion of removal of symptoms. It is the resolution of the causes of the disease. Healing is returning to a state of alignment with our Higher Self, with the Source, or true way of being. It is balance.

The main principle in reiki practice, as in other disciplines that include energy healing, is that life force nourishes the organs and cells and when it is blocked or disrupted, the body ceases to function well. Reiki practitioners facilitate the process of healing through the *laying on of the hands*. The Universal energy enhances the flow of the person's energy or KI.

We find a similar concept in Traditional Chinese Medicine, in some healing methods used in Japan, in Pranic healing, esoteric healing and Ayurvedic medicine.

Practitioners can use their hands over the aura or touch the person receiving the treatment to help unblock the flow of ki, inducing a deep state of relaxation and promoting the restoration of balance.

The effect of reiki is not limited to the physical. Because we are multidimensional bodies, reiki affects all levels of our being. There is a parallelism between communication between organs and the connection between subtle bodies. When the biomolecular communication between organs is impeded, the body cannot function optimally. We cannot be happy or healthy when our level of awareness restricts the communication between the different dimensions of our body (physical, emotional, mental and spiritual).

Reiki effects are lasting if the person who receives the treatment makes a commitment to transform what was causing disease. If the cause persists, the person will soon lose balance again.

A reiki session is a pleasant and relaxing experience, of great help in emergencies as it alleviates pain, sedates and can even stop bleeding. Reiki induces change, but eliminating the causes of disease is each person's responsibility. If I suffer from frequent headaches, reiki can alleviate the pain, but if they are caused, for example, by the excess caffeine that I put in my body, the ache will come back until I introduce changes in my habits.

Some of the people who want to experience reiki and come to the circles or for a session, are expecting a miracle, and miracles occur, but only when the person is ready. Being ready usually means that the person changes perspective, that they take a leap and change their attitude, their thoughts, and their behaviors. Little changes will accumulate until they produce a qualitative change. Then the person will be open to the energies that surround her and the miracle will happen. Reiki and the tutoring of the master play an important role in the process.

For many years, I have invited family, friends and acquaintances, in particular reiki students, to examine their beliefs on health and illness and how those beliefs frame their relationship with their bodies. I have insisted that a miracle is a change in perspective/perception and that they have to work for the miracle to occur.

Change doesn't happen if the person is not ready, if the person doesn't have the willpower to transform their lives or accept what cannot be changed. On many occasions we express the desire to change habits or attitudes (take New Year's resolutions) but we are so conditioned that we become prisoners of our habits. On the other hand, I have seen people who express skepticism and resistance to change and still, in the deepest levels of their consciousness, they are already impelling the changes they need to move forward.

Founder, Mikao Usui (1865 – 1926)

I will summarize the history of reiki, based on what I learned from my masters and what I have researched since I started to practice reiki in 1994. My first master was Maria Adelina Sastre, from Spain, who belongs to the *Reiki Academia del Mediterráneo* (RAM). She initiated me in Cartagena to reiki levels I, II and III – levels I and II in her school were the equivalent to what in the traditional school is presented as level II. Beatriz Eugenia Cárdenas was the Colombian master who initiated me to the reiki master level.

To tell you the history of reiki, I will include some information from sources that I consider reliable, like reiki masters William Rand and Frank Arjava Peter. At the end of the book you will find a list with references if you want to read further about the topic.

History and legend blend to create the story about Usui that you will find in most reiki books. Maureen J. Kelly in *Reiki y el Buda de la Sanación, (2000, Reiki and the Healing Buddha)*, says that there are many layers of symbolic and metaphysical meanings in the story to guide the student in the spiritual path. Usui was a Buddhist and in his religion, stories usually have hidden meanings.

It has been told that Mikao Usui was a teacher in a Christian school near Kyoto and that his students challenged him to explain how to do hands-on healings. Usui took the challenge and came to the United States to study theology at the University of Chicago. After seven years of study he went back to Japan.

Rand has found that there is no evidence in the records that Usui attended the Chicago University. Some authors say Usui was a Buddhist monk, others say he was not, and that he might have studied Christian religions, but there is no evidence that he was a Christian.

The story continues to tell that Usui traveled around inquiring about hands-on healing but he was unable to find the answers he was looking for. Hands-on healing seemed to have been "discontinued" in the world since the times of Jesus.

Some authors say that during a cholera epidemic, Mikao Usui fell sick and had a near-death experience. He understood that his purpose in life was to devise a healing method that could benefit humanity, a method that would use both modern and traditional knowledge. Apparently, Usui belonged to a Buddhist family from the Tendai sect that didn't accept this revelation. Estranged from his family and religious community, he sought the guidance of Watanabee, a Shingon Bonze (monk), with whom he studied Buddhist esoteric healing methods.

Most recent recounts of Usui's story tell us that during the late 1880s he was exposed to certain manuscripts, which turned out to contain the healing methods he had sought for so many years.

Among the manuscripts was the *tantra*[47] of the *Lightning Flash*, "the secret transmission for healing all illnesses of body, speech and mind." This tantra provided the information that he had been looking for and presented a comprehensive healing method. It is said that the manuscript was written in old Japanese with notations in Chinese and Sanskrit, languages that Usui studied to better understand the writings.

It seems that the teachings contained in the tantra belonged to an ancient mystic branch of Hinduism, and that the healing method was derived from esoteric Buddhism as it was practiced in Tibet, including the method to invoke a superior being that could concede the gift of healing. Usui practiced on his own until he felt he could go no further. Maybe some information had been lost in translation.

The monastery's abbot recommended a retreat.

After 21 days fasting and meditating in the sacred mount Kurama, Usui felt as if he had been stuck by lightning in the mid forehead and fell unconscious. When he awoke, he was refreshed and rejuvenated and felt that the universe had penetrated his mind and body resonating with his own inner divinity. He felt at one with the universe.

Going down the mountain and after he injured his foot and relieved the pain and the bleeding by the laying on of his hand on the injury, Usui realized that he had been giving the gift of healing.

With the characteristic discipline of a Buddhist and after seven years of intensive practice, Usui created a school and a treatment center and initiated others to reiki. In his early experience with reiki, he worked in a town of beggars and performed many healings. Notwithstanding, he realized that beggars didn't seem happy or grateful afterwards and would come back to beg because they found it hard to work for a living. Usui left the beggar's district and concluded that they didn't value his healing because it cost nothing and they were not ready to take responsibility over their lives or their health.

Usui also understood that he was somehow reinforcing the

---

[47] Tantra: Hindu or Buddhist scriptures. Sutra: precepts that summarize Vedic teachings. Veda: Hindu sacred teachings.

patterns that characterized a beggar: receiving without giving back anything. Was he interfering with their path?

His disposition as an apprentice and his humility are exemplary for us and a motivation to think about our tendency to interfere with other people's processes. We may unconsciously try to control others through what we deem generosity and inclination to service.

After seven years, Usui started to initiate others into the practice of reiki. In 1922, he founded the reiki society, called Usui Reiki Ryoho Gakkai. Although it seems this society dispersed by the end of the Second World War, Arjava found that it is still in existence in Japan. Arjava has been able to make contact with practitioners who studied with Usui and he recovered and published an instruction manual by Usui that he published a few years ago.

Just before noon, on September 1, 1923, a devastating earthquake affected the densely populated, modern industrial cities of Tokyo and Yokohama. Deaths were estimated at nearly 100,000, with an additional 40,000 missing. Thousands of homes and restaurants caught fire and many buildings were demolished. There were flammable materials in the industrial plants and a munitions factory exploded, feeding the flames at a pace that the firefighters could not keep up with.

Usui and his students responded to the catastrophic event, offering reiki to victims. Soon his fame spread and many healers and physicians started to request to be initiated to reiki. Usui received a very high award (a Kun San To, the equivalent to an honorary doctorate) from the emperor.

Naval doctor Chujiro Hayashi was among Usui's non-Buddhist students. He was a Methodist Christian with very strong beliefs, and maybe not very open to the esoteric nature of Usui's teaching. Hayashi eventually opened his own clinic in Kyoto and replaced some of the format of Usui teachings by creating a system of 'degrees' and developing his own set of hand positions.

During World War II, Hawayo Takata, who had been healed in Hayashi's clinic several years earlier, was invited to Japan to be initiated to the master level. She introduced reiki to North America in the seventies. She trained 22 masters. After she died in 1980, her granddaughter Phyllis Furomoto created the Reiki Alliance to continue Takata's tradition. However, today, hundreds of reiki schools teach reiki around the globe, each one introducing their own version of the history, their own practice protocols and their own curriculum. It seems clear that many of the original teachings were lost.

Notwithstanding their differences, reiki masters agree that reiki can be used either as a tool for self-enhancement or as an instrument to facilitate healing processes in oneself or others. Different from other energy modalities, reiki is tradition that requires an *attunement*. Attunements by the master are thought to open the practitioner's *channels*, initiating the student to reiki. After the initiation, the practitioner is ready to funnel the universal energy each time she sets up the intention, and to apply it to others or self for the rest of her life.

Reiki energy is considered neutral and is always present in the practitioner's body. Any person can receive or be initiated as a reiki practitioner independent of beliefs, religion, culture, age, lifestyle or experience.

Reiki precepts

The underlying unifying principle of major religions and spiritual practices is love. We are talking about a kind of love that is unconditional and impersonal, that doesn't generate attachments and doesn't ask anything in return.

Love is the opposite of fear and fear is the origin of preoccupation. Love is the opposite of anger and anger is the origin of hate and separation among human beings.

We respond by either fighting against a perceived foe or running away when we feel threatened. I see that the *fight or flight response*, which is instinctual, has its mental and emotional counterparts in fear and anger. But in the same way that the body's adrenaline surge is over after a few minutes, our fear and anger should subside as soon as we evaluate what was perceived as a threatening situation and feel safe again.

However, we live in a world that stimulates fear and anger. Even though the odds of being directly affected by such things as terrorism, avian flu or natural disasters are statistically low, they have made us feel and think that we are never safe and we live by that. If an extraterrestrial visited the planet and watched the news for a few minutes, what would his idea of the world be? Aren't there wars and killing rampages, foes and hate everywhere? Financial safety, physical safety and national safety have become the main concerns in a world where we are constantly dreading death. Fear gets in the way of trust and love and become a rational for restricting civil rights and freedom.

Besides promoting worship to a superior being, a religion compiles certain habits and makes them norms that aim at promoting harmonious relationships among human beings, humans with nature and man with self. The sacred books – Old Testament, Torah, Koran – are compilations of oral tradition, word of mouth stories passed down from generation to generation. Rules of behavior, of life together, and hygiene, included in the books, ensured in its moment the survival of the people and the endurance of the tribe.

Psychoanalyst Carl Jung affirmed that religions are psychotherapeutic systems and I would add that they are so when people practice what they preach.

Reiki is not a religion and doesn't have a dogma or doctrine, but Mikao Usui simplified for his students a set of precepts that seem to have their origin in the *tantra* of the *Lightning Flash*. The precepts are just life norms that, as Usui says, invite happiness.

Some reiki authors sustain that reiki precepts or principles were originally issued by Emperor Meiji who ruled Japan in the times of Usui. Notwithstanding, after having researched the topic, I found no mention in history books or Web sites about this. On the contrary, I found that the emperor embraced Shinto, an ancient religion that had no sacred books, no dogma and no other precept that truthfulness.

Because it seems that Usui taught reiki for enlightenment to his Buddhists disciples (who have already taken *refuge*) and a simpler version of the method for other people, I think that the principles that we have gotten from Takata are a simplification of the principles found in the *Lotus Sutra* and taught by Ippen. These were precepts originated in Japanese Buddhism that were popular in Japan beginning in the ninth century.

Following the precepts might be a first step on the path for spiritual healing. They refer to moral development, which implies not to harm oneself or others in thought, word or action. Repeating the precepts during reiki meditation becomes a tool that helps to silence the mind and achieve deeper levels of concentration, which are necessary to develop intuition.

Moral development, mind control and achievement of wisdom are simultaneous processes, but their depth depends on the individual's spiritual development.

Buddhists have preached the middle way, a way of life where precepts are not accepted as an imposition, but studied and understood before they are incorporated and practiced.

We human beings tend to go from one extreme to the other. When we try to amend an excess, we commit the opposite excess. We starve to lose the weight that we gained by eating too much. We produce surpluses that we cannot use to remediate scarcity. We have gone from slavery to licentiousness, from control to negligence.

The precepts are a guide. They don't have to be adopted out of fear of being punished; we follow them because they have intrinsic value.

If someone decides to follow a spiritual path, it is normal to take up discipline not with mortification, but in the search to gain knowledge and purify our hearts.

A spiritual path doesn't need to include worship or prayer or punishment. Prayer might come naturally in our relationship with the source, the supreme, the universe, everything that is... God.

Precepts don't need to be memorized. I recommend studying them to see how they fit with the principles you have already adopted. To build the foundation that guides our life is our own responsibility. Severity does not exist in the middle path. We follow the path seeking to live in harmony with the universe. The person who achieves this harmony will feel protected amidst calamities because from the perspective of the soul, every life event is perfect as it is and can be assumed to be a learning experience in our personal evolution.

In table 7, we offer the principles contained in the manual that Usui delivered to his students, as it was published in *The Legacy of Dr. Usui* by Frank Arjava (reproduced with Arjava's permission).

Initiation to Reiki

Reiki can be used as an instrument for self-enhancement and also as a resource to support the healing process of another person. One of its peculiarities is that reiki practitioners can do the hands-on healing on themselves.

Reiki can be described or explained as the use of universal energy, which is intelligent, to achieve the effect needed by the body. There is no energy "exchange," and there is no manipulation of the energy of the patient. Reiki uses the intelligence of the universe to stimulate the energy of the multidimensional body. If a symptom is not alleviated by reiki, it might be an alarm to which the patient needs to listen.

Table 7. Reiki Principles (Japanese and English)

Shoufuku no hihoo	The Secret Method Of Inviting Happiness
Manbyo no ley–yaku	*The wonderful medicine for all diseases (of the body and the soul)*
Kyo dake wa	*Just today*
1. *Okuru–na* 2. *Shimpai suna* 3. *Kansha shite* 4. *Goo hage me* 5. *Hito ni shinsetsu ni*	1. *Don't get angry* 2. *Don't worry* 3. *Show appreciation* 4. *Work hard (on yourself)* 5. *Be kind to others*
Asa yuu gassho shite, kokoro ni nenji, kuchi ni tonaeyo	*Mornings and evenings, sit in the gassho[48] position and repeat these words out loud and in your hearts*
Shin shin kaisen, Usui Reiki Ryoho	*(For the) improvement of body and soul, Usui reiki Ryoho*
Chosso Usui Mikao	*The Founder, Mikao Usui*

My neighbor had a heavy object fall on her toes and I administered reiki for several minutes until the pain was gone. Notwithstanding, an hour or so later the pain was back and one of the toes was swollen. She had fractured one of the tiny bones in a toe and pain was alerting her that she had a serious trauma. This allowed her to take necessary action to immobilize her foot and grant prompt recovery.

One morning I fractured a toe myself and didn't realize it until I put my shoes on. I treated myself with reiki for probably two hours, after which the pain and the inflammation were almost completely gone. My toe was black the following day and painful if I touched it or used shoes, so I rested and elevated the foot. I decided not to have a cast on my foot,

---

[48] The hands in front of the chest like in a prayer, a little above the heart.

but I continued to use reiki, and pain helped me to connect with my body and monitor the healing process. I avoided walking and using shoes until the toe completely healed.

A few days after the injury I attended a meeting where a friend who is a reiki practitioner laid her hand on my foot. I was speaking when I started to feel funny, as if I had drunk a glass of wine (I don't drink alcohol). She was only touching my foot but reiki had a global effect on me.

After a trauma, we usually use reiki over the part that is affected but the whole body receives the benefit.

The initiation to reiki can be understood both as a rite-of-passage ceremony and as a transformation in which the initiate is 'reborn' into a new role.

I cannot provide a scientific explanation of what happens during the few minutes that the attunement lasts. By tracing certain symbols over the aura, the master opens the student's *channels* (whatever they are) or aligns the chakras, as other people explain it. Most students experience peace and a feeling of connection with everything that is. Others have mystical experiences and even other experience almost nothing. Still, after that moment, all students report a difference in the quality of their touch, a sensation in the hands, which will be present from then on and increase with practice. We describe the initiation as the moment in which the student becomes a receptor/transmitter of universal energy.

Reiki energy is neutral (not positive, not negative) and always has an effect independently of the beliefs, religion, culture, age, studies or experience. The way it is studied in North America, reiki is learned in three stages or levels.

First Degree. It is focused on the physical level. Practitioners start a process of self-awareness working on the physical level and learning to administer self-treatments, as well as to treat others.

Second Degree. Focused on emotional and mental dimensions. Three of the five reiki symbols are learned and also distance healing.

Third Degree (Master level). The practitioner is initiated to mastery, which is understood as working to gain awareness and command over oneself, not others. There is not a single course where you can gain mastery. Achieving mastery is a long and committed process where we are truthful and stop denying our dark side. Two more symbols are learned and the practitioner might afterwards proceed to be trained to teach others. This may require the student to attend several trainings as an assistant before being ready to be a teacher. Each master has different requirements to pass on the master degree.

As universal energy is all intelligent and cannot be constrained within time or space it doesn't seem that there is a specific time to receive the initiations. The student intuitively knows when she is ready for the next degree and the universe agrees by facilitating the teacher, the time and the resources.

Most master/teachers are eager seekers in a quest for knowledge and explore many disciplines that they add to the reiki class (some do it to gain a marketing advantage or add credibility). Therefore, there is no one single way to teach or deliver the courses. I don't believe it is necessary.

Reiki Seminars

You cannot learn Reiki in a book or online (even though some people offer it this way). If you want to learn Reiki, you need to attend a seminar.

I usually begin a class by putting forward a couple of arguments to support the idea that you cannot teach reiki. This of course generates certain uneasiness. After all, these students have registered in the class to learn reiki, that's the expectation they have of my seminar.

However, reiki is not a doctrine that you can pass on, nor a technique in which you can be trained. Of course my classes include a segment to practice sets of hands positions, meditation and relaxation techniques, information about the energetic structure of the body according to different beliefs and some advice to support the *inner healer*. Still, that is not what reiki is about.

It would be too much to pretend that information shared during a few hours would suffice for people to learn how to meditate or do visualizations, or that in just a weekend-long class the students will do an about turn and leave behind decades of old habits or beliefs that they had never questioned before. However... in a few cases, this has happened.

We have all attended lectures, classes and workshops and we may agree with the following expression:

We learn 20 percent of what we read; 30 percent of what we hear; 40 percent of what we see and hear and 90 percent of what we teach someone.

So, even being extremely optimistic, I don't expect that the students will retain a lot of what they hear from me, unless I am answering their specific question or matching their own experiences. Most of what is said will come out as fast as it went in, but at least some of the questions the student brought to the class will be answered. Practice is something else. After you touch and experience the energy, you've got something that you won't forget and that you will be able to use with others. As we mentioned before, true learning implies a change in our vision, our attitudes or our behavior.

My second point, which I have defended for years as an educator, is that *nobody teaches anybody anything*. The knowledge lies within the community, and not within the expert, said psychotherapist Carl Rogers. Knowledge is the result of individual and collective experience. Therefore, the teacher's role, the master's role, is to help in the digging of a well, until they find where the knowledge lies. No less, no more. Once the student finds the source of knowledge, she can come back to it as often as she wants. If the knowledge is missing, only the learner can find it.

Teachers can contribute to creating the kind of environment that is conducive to learning. That's why the teacher tells stories, shares her own experiences, listens and cares for the students, promotes laughter and introduces exercises that are thought-provoking. Teachers can play an inspiring role.

Our behavior as teachers sets an example, sometimes in a positive way and sometimes by showing what should not be done. Often, what is most important in our discourse is to support another person to *de-construct* limiting behaviors, attitudes and beliefs.

I like to structure the classes as spiritual retreats. They become a safe and confidential space where students can connect with their inner selves, where all the participants become engaged in each other's processes.

Frequently, as we practice the hands-on protocols, students show concern about their *performance*. We have gotten so used to evaluating, rating what we do as correct/incorrect, good/bad, pretty/ugly, that we forget that each one has a unique way of doing things and we should not compare ourselves or others to an external parameter. The beautiful poem *Desiderata* says,

> *If you compare yourself with others, you may become vain or bitter; for always there will be greater and lesser persons than yourself.*

Besides, the map is not the territory. It's not the way you place the hands what makes reiki work. It's not memorizing the exercises recommended to raise the vibracional frequency or the information shared what makes a practitioner.

On occasion, something that the master or one of the students says or does, upsets another student. Great! I recommend the student then to follow the lead, to go beyond the threshold. This uneasiness might be reflecting resistance to change that the student is not yet aware of. Until the student gains understanding of the issue, and processes and overcomes the resistance, the student might avoid the teacher.

We are full of excuses and intellectualizations to explain our moods and resistances. We might choose to adopt the perspective of the victim instead of being an apprentice. But only apprentices can become warriors and *Men of knowledge* (to lend Don Juan's terms in Carlos Castaneda's books).

A reiki master can help students rekindle the inner apprentice. For this, the master should be aware that students come with a rich baggage full of previous experiences and also be aware of her own authoritarian tendencies.

When life makes me play the role of a mentor, I am grateful to have the opportunity to witness another person's discoveries, their inner treasures, the wealth of knowledge they were not aware of and their skills.

Those who know me well know of my tendency to not give straight answers to closed questions and also that I often answer with another question. If I am asked for advice, I may respond with questions like these:

What alternatives do you see to that situation?

If the solution of the problem totally depended on you, how would you start solving it?

If you were to share your experience and knowledge, how would you design a seminar on that topic?

How would your life be if it were not dominated by that thought?

These questions might seem simple but they can turn into a philosophical framework, an outline of a program and maybe, in the end, in a seminar that the former student can offer successfully. Or in key changes in a couple's relationship. Or in the development of an initiative that had not been considered previously.

When we think of problems not in terms of grievances, but in terms of finding solutions, we might discover paths that we never saw before.

The advantage of adult students in general and reiki students in particular, is that they seek the teacher voluntarily. This makes them open to the learning process, which is not limited to the classroom and does not end with a diploma. It is an ongoing process.

The attunements during the initiation have a similar effect to when you revive a battery to jump-start a car.

The initiation is an intimate, unique and mystic moment; an alchemic moment. It is said that the main goal of the alchemists was the transmutation of *impure* or *ignoble* metals like lead into *pure and noble* metals like gold and silver. Esoteric schools consider this a metaphor for the transformation of the spirit. Gold has been greatly appreciated in different cultures probably because it's a shiny and hard yet ductile, malleable and stainless metal. On the other side, lead is a soft, heavy, metal that tarnishes to dull gray when exposed to air. It has a low melting point. The task of the spirit would then be to transform our lead into gold.

The alchemists had a dynamic vision of the world and believed in change as the constant element in life.

The Kybalion states that,

> *Mind (as well as metals and elements) may be transmuted, from state to state; degree to degree; condition to condition; pole to pole; vibration to vibration.*

When someone decides to take a reiki class, they are opening a door that might lead to *transmutation*. The students have chosen, even if not totally consciously, to walk a new path, which requires certain information, and a good gear of supplies and resources.

That's what the classes are for. Mikao Usui's legacy is this marvelous yet simple healing method. He wanted everybody to have access to it. Classes are delivered so that people can have access to reiki. Seminars are most of all a stage where students are inspired to use this invaluable tool to start a process of transformation. When the vibrational rate of the student raises, he can resonate with the universal energy and use this energy in his own benefit or to support others.

Reiki practice

When in 1982 Phyllis Furomoto met in Hawaii with the 22 masters her grandmother Takata had initiated to reiki, she found that they had not been taught the exact same things. Not even the symbols taught were exactly alike. "Takata's unique method of teaching was a source of great upset. We did not understand the uniqueness and came to it with our Western notion of uniformity and standardization," said Carell Ann Farmer, one of the masters present in the gathering, in a public letter published by William Rand.

Today there are probably 30 or more reiki schools in North America, all teaching variations of Usui's system. Different teachers of the same school tailor the classes according to their own experiences, and thus there is no one single curriculum for teaching reiki. Most books contain illustrations that help the student remember the positions of the hands and the positions also differ from book to book.

When administering reiki, the practitioners touch the patient with their hands cupped to concentrate the energy that is coming through a secondary chakra in the center of the palm. In some occasions the treatment is administered with one or two fingers, like when using the pinky to treat an ear ache.

In one occasion I offered a session to a friend who had taken reiki I. She was convalescing from a laparoscopy and her abdomen was inflamed and sore. After a few minutes with my hands on her abdomen, both her skin and my hands turned very hot, but this is a common sensation when you deliver reiki to an inflamed area. It is a sensation that invites you to extend the time you leave the hands on that spot. After a few minutes, she felt alleviated, but the most interesting thing was that my own hand was relieved from a pain I had been experiencing for several days.

Although reiki practitioners usually experience the benefit of reiki while giving a treatment, in this occasion it was clear to me that I was receiving energy through my friend's abdomen. I would say that in reiki we truly learn that *giving is how we receive.*

Usui believed that the whole body emitted energy, but especially the hands, the eyes and the mouth. He laid his hands on the aura or directly on the skin, sometimes using massage or tapping. He would also focus his sight on a problematic area or blow over it.

I tend to go by intuition and don't always follow the hands protocol that I learned, but I find that it gives new students a good starting point while they develop their intuition.

When providing a treatment, I mingle what I have learned in all three levels. I use symbols, visualization, and tapping, anything that seems useful in the moment, because I have learned to trust my intuition, which I understand as the result of my connection with the universal energy. I would say that the energy "informs me." That's why I recommend students to follow their impulse to take the hands to certain spots in the body and leave them there until it feels like enough. I usually start a treatment at the head and then I forgo the protocol and work with the client according to what I feel the energy is dictating me to do or say.

If practice makes masters, in reiki you have to give treatments to become a master. A treatment is not about applying a technique but rather establishing a connection with that intangible intelligent energy that guides our doing. I would describe the experience as a connection with God, with everything that is. While giving a session, many practitioners experience the presence of what some deem guides, others call masters and still others, angels. People often feel a presence in the room and feel more than one pair of hands touching them. The experience is always very calming and soothing, unless the receiver rejects the treatment for some personal reason.

After each initiation, from the first to the third level, masters recommend the students practice on themselves for at least three weeks before administering to others. It only takes a few minutes to care for ourselves; it becomes a healthy pause that in time teaches us to listen to the body and respect its needs. Listening to the body leads to a rhythmic life. The self-treatment helps the student not only review and memorize the positions of the hands and experience the energy, it also seems to trigger a detoxification process (maybe energetic but expressed in the different dimensions of the body with physical symptoms and sometimes emotional upheaval).

In my experience, a daily treatment recharges vital force and stimulates our spiritual growth, increasing our awareness. This awareness helps us be in touch with our true feelings, leads us to acknowledge our secret motives and promotes our personal growth. This is a kind of knowledge that cannot be learned in a class or through a book.

After the first three weeks, the student can start administering treatments to other people, animals or plants. A full treatment for a person should include four one-hour sessions, leaving the hands for three to five minutes in each position. In chronic ailments, the treatment can be prolonged for weeks or even months. It is important to remember always that the goal is not necessarily the disappearance of symptoms but achieving wholeness.

In emergencies, hands can be left on the injured or painful spot for as long as we deem necessary. Research has shown that reiki helps to stop bleeding and accelerates the healing of wounds.

Reiki effects do not depend on our will or the will of those who receive the treatment, but on the universal laws, always seeking to restore and maintain harmony. "Miraculous cures" seem to happen only after people have elevated their consciousness, which makes them reconsider their lifestyle, habits and relationships. And as said before, I believe that a miracle comes from a change in perspective.

Reiki is also applied to people with terminal illnesses and, because it has an effect in each one of our different bodies or dimensions, it helps the patient accept the transition. It sedates and calms the patient.

Whenever possible, it is preferable to administer full body treatments plus a special treatment over the problematic areas. Ideally, the session is given in a tranquil atmosphere. Many practitioners use incense, candles and relaxation music, according to patient's preferences. When I start a treatment, I enter a meditative state and when I finish, I feel refreshed and energized.

A common question during reiki classes is whether we can be affected by the energy of the receiver or if we can "catch" their symptoms. I know that pranic healers place great attention to the cleansing of the aura and the chakras of the person they are treating and of self. From what I have heard and read from other reiki masters, they seem to agree that because we are working with universal energy, this energy will protect us. I have never had a negative experience administering reiki, but other modalities of energy work have not been always positive for me. I don't have an explanation for this.

In the personal process that starts with the first initiation, I recommend starting a private journal in which to register the sensations and experiences related to our practice. This helps to reinforce our growth and raise our consciousness, at the same time that we can follow up on our spiritual progress.

The master

Experience seems to confirm that we end up in groups according to certain affinities; we are attracted, or we attract to us, those who are on a similar path (some might say, we attract those who have similar vibratory frequencies), and masters who have the answers to our queries.

The concept of mastery is two-fold. It refers both to the art of educating others and to the dexterity achieved in the practice of an art, a technique or a science.

I have asked myself and heard others ask, how do you choose a master? In today's world, you find "masters" everywhere. If we truly have the attitude of the apprentice we will make masters of all who surround us and will learn from them whatever we need to move ahead in our search.

If the master appears when the student is ready, I see it not as a coincidence but as a consequence. When we are ready for the knowledge, we get in tune with what surrounds us, which facilitates the process of perceiving the answers.

In the Western world you rarely see a student following a master for a lifetime. Most masters are momentary. It is natural to create bonds with whoever indicated a path to follow when we were lost, but a true spiritual master doesn't aim at controlling us, gaining followers or generating dependencies. On the contrary, a true master encourages autonomy and empowers those looking for her coaching.

History has recorded different kinds of masters throughout time. Most of them were not simply instructors or tutors. Besides wisdom, they had achieved command in their field, like for instance, artists of great skill (Michel Angelo, Leonardo), and those who taught us how to live exemplary lives, like Jesus. There have also been writers (Flaubert, for example) whose patience and perseverance earned them a page in the annals of time. And many anonymous carpenters, masons, artisans have left their footprints in their masterful work, and/or have passed on their art to the next generation.

We have all heard of masters who excelled in their trade and kept with great zeal the secrets of their trade. We have heard of sages like Socrates who used their inquisitive minds in the service of their disciples. Or scholars whose rigorous discipline merited them high positions in society. Or gurus who stimulated their disciples' minds with unsolvable questions. And leaders who could see beyond most mortals of their time.

Nowadays, you can get a diploma in mostly anything on Internet without much effort, with just a few bucks. And it makes it difficult to discern between the sages that the universe brings daily to our lives and those who just have titles.

If you are truly after gaining wisdom, it is fundamental that you cultivate a disposition to learn. The master will then show up.

168

In reiki, we look for a master who opens the channels through which universal energy is funneled. Also, we are invited to look inside ourselves and find our inner master who will guide us through the different challenges presented in our development process.

On becoming a healer

Generally, when someone hears the term *healer*, he imagines someone capable of performing miracles, massive healing, spectacular deeds. This popular vision has, of course, a historic root, because it is mainly from unique masters like Jesus and his disciples from whom we know about healing by *laying on of the hands*.

Therefore, it is thought that healing requires supernatural powers and is not accessible to common human beings. Some of these ideas are reinforced by healers who set up shows where people are *miraculously healed*. But we need beware of *charlatans* and understand that people with certain personality disorders may all of a sudden fall under the healer's suggestions and "heal" from ailments that have a strong emotional component.

Extra-Sensory Perception phenomena (ESP) like clairvoyance and telepathy are also considered by some as good proof of the extraordinary powers of a so-called healer. However you don't need to be a spiritually evolved person to develop ESP and this idea of the healer as someone with extra-ordinary powers creates a ditch that separates common people from healers. It also makes people believe that they are just passive recipients of the healing, mere spectators or witnesses.

I believe that as human beings we all have the potential to contribute to another person's healing process. We affect others even when we are unaware of it but we can choose to significantly touch others. We may choose to validate or destroy, to reinforce or minimize, to accept or reject. We can love or ignore, trust or fear.

Many people believe that to choose a healer they must look someone who leads a special kind of life. It is of course important to look at the coherence of the one who calls him or herself a healer. Notwithstanding, the healer is also an apprentice on the path to evolving and discovering her own true essence. We are eternal souls experiencing limited human lives.

There are even those who think healers should display certain credentials because knowledge is what is most appreciated these days. If the healer is a reiki practitioner, some people want to know about the practitioner's lineage or school, for example. In my view, this is a vestige of medieval times and corresponds to an elitist pretension to keep knowledge accessible only to a select group of people.

Reiki is such a simple method that it helps us redefine the concept of *healer*. Usui, according to the manual published by Frank Arjava Peter, alludes to the Japanese tradition in which, when somebody discovered a secret law, they kept it secret to pass it on to their heirs or disciples and maintain it safe from intruders. However, in these times, Usui says, and he is talking at the beginning of the 20th century, "happiness of humanity is based on working together and the desire for social progress." Usui wanted to make his method available to the public for the well-being of humanity. "In this way," he says, "a great many people will experience the blessing of the divine."

He adds that reiki "is something completely original and cannot be compared with any (spiritual) path in the world."

With reiki, we become healers, intermediaries of the universal energy. It is not necessary to embrace certain beliefs or norms or to follow certain lifestyles in order to practice reiki. This healing method can be adopted by the practitioner as a path that each one traverses at their own pace. The point of departure and the journey are unique to the traveler. Some practitioners limit the use of reiki to the physical body and the body is their route. Others use it only for spiritual reasons and still others find it useful in all circumstances.

By channeling reiki energy and practicing reiki principles on a daily basis, a person may learn to align her will to the will of the universe. Diving into a process of self-discovery, that person will see her intuition and empathy developing at par with her capacity to be present.

Common intuition is not otherworldly. Our senses pick up millions of impressions per second but we are only aware of a few. It would be too overwhelming for our brain to process all of them at one time. Some of these stimuli are vibrations detected by our subtle bodies and a number of physical stimuli are detected by our senses without our awareness, which explains why people under hypnosis may remember what they cannot when they are in their normal state of consciousness. Our brains are far more efficient in the processing of information than any computer built by humans, and can arrive at conclusions without us having to deliberately work on a question. We perceive the inferences made by our mind as intuitions or precognitions, which in most cases are good guesses because they didn't pass through the strainer of our prejudices and beliefs.

Some authors sustain that all children are born with psychic capabilities that atrophy as they grow, mainly due to the fact that adults repress the children's stories, considering them either lies or the product of rich imagination.

Psychic Edgar Cayce, by the end of last century, said that we could all cultivate and use those intuitive faculties that are latent in our souls.

Some people develop great intuition by disciplined practice, some as part of a spiritual path, some others because of the need to assert control over others or over certain situations. Nobody can measure the spiritual development of another person based on the development of their intuition or ESP capabilities. Spiritual development, in my view, shows as expanded awareness of self and others and the capacity to love unconditionally and feel compassion.

A healer is then a therapist, in the sense in which we have used the word, an intermediary in the process of healing. As we gain the skill to recognize that our actions and omissions affect the course of things, we will assume our cosmic responsibility and learn to be therapeutic in all circumstances. But therapeutic doesn't mean restricted to the sphere of the therapist. It means that we affect each other, something we must be always conscious of, because what we say, what we do and what we omit can contribute or not to healing another person. And this is of concern especially for those of us who have chosen to become facilitators of other people's healing process.

Know thyself

The therapeutic work starts by learning to accept and value ourselves and then to accept and value others. It is important that we stop making comparisons in which we grow taller or smaller depending on how much we value the person with whom we are comparing ourselves. The habit of making comparisons is born from the tendency to appreciate not what we are, but what we have achieved in material terms (money, fame, social status). We must understand that our achievements stem from our strengths, but our achievements are not our strengths.

In the process of self-discovery we need to examine what made us take the path of a healer and why did we choose professions in which we take care of others. It is important to gain awareness about our intimate, unconscious purposes and rationales.

Fifteen years ago, I read for the first time *The Drama of the Gifted Child* by psychoanalyst Alice Miller. Her ideas encouraged me to look at the connection between the professions we choose and the metaphors that define our lives, an allegory inspired perhaps by our main need, or by a problem that remains unresolved.

Here are a few examples of those possible metaphors:

√ attorneys whose lives are focused on prosecuting or defending their loved ones;

√ judges dedicated to a quest for justice, who won't stop judging other human beings;

√ doctors who have played the healer's role in their families or become obsessed about their own health;

√ people who have chosen information technology as their profession, whose main challenge is communicating with others;

√ sales people for whom life is like an immense sales department and who expect to get "commission" from every situation they deal with;

√ artists and writers who depict through their works an order that doesn't exist in their own lives;

√ traders devoted to sell their ideas to others;

√ architects constructing and de-constructing their own lives or the lives or their close ones.

And there is of course, the *wounded healer*.

We live in a culture characterized by codependent roles rather than healthy interdependency. Almost daily we form some kind of symbiotic savior/saved dyad that expresses in our husband and wife, mother and child, therapist and client, doctor and patient relationships. The rescuer stimulates dependency, not consciously, by offering to solve a problem, delivering prescriptions, providing advice or support. The saved one stimulates (not consciously) the heroism of the other by maintaining their vulnerability, incapacity, need or illness. These roles provide compensation for the validation that the person needs but create a great risk: their world can crumble to pieces at any moment, because it depends on something external that cannot be controlled. This explains why, in codependent relationships, we see a tendency to control and manipulate the other. We manipulate through advice and though pain. We control with *love* and by displaying weakness.

Nonetheless, I feel that we humans are moving towards some exciting new ground, full of possibilities. Our first steps towards emancipation look like egocentrism and indolence. But really, we are learning to take responsibility for our own businesses, at the same time that we recognize how important it is that each one take responsibility for themselves.

As a complement, it is necessary that our actions be guided by love. We don't want to be indifferent, we are learning compassion, (which means to share passion, or to love with wisdom), and we are leaning to become empathetic. Compassion is something we can learn to make part of our daily lives. Instead of judgments, we can practice seeing the world from the other person's shoes.

I accepted an invitation to participate in an exercise that surprised me. I found myself making value judgments automatically, when I thought I was no longer judgmental. The goal was simple; I had to recognize in each passerby the perfect being that exists within each one of us. I had to bless every person I ran into during my morning walk.

I realized that certain people could hardly elicit a bit of kindness on my part, just because, for example, they would not return my smile. If I couldn't visualize their potential perfection, I could not relate to that potential perfection inside of me either. It made me understand how many times I am hard on myself and how my perception of myself and others is many times tinted by my mood.

From my colleague Carmen Escallón, who inspired my with her enthusiasm to explore the holistic path that I still follow, I learned, among other things, a new definition of violence: *Violence is not to allow the other to exist in her own terms.*

We contribute to the healing of other people when we validate, affirm, accept, encourage and provide space for their expression. Opposite options are not loving choices, and are also violent: to judge, control or punish with abandonment.

Reiki implies the commitment to evolve and discover our true essence. That's why pondering *where are we* helps to orient us and to develop certain qualities that will allow us to support others.

One of Usui's precepts is to work honestly with ourselves. When we refine our character by developing those qualities that characterize a healer, when we raise our awareness and recover our lost integrity, when we heal, we can then expand our spiritual world.

Let's also be down-to-earth. Our destiny is not faultlessness but totality, wholeness. Spirituality is not about the implausible elimination of all of our blemishes to become truly pious, but a sincere effort to stop denying our dark side.

If we promptly acknowledge and process the ugly contradictions that dwell in all of us, without denial or self-deception, then we will make progress in our task of supporting the healing of ourselves, others and the planet. By being aware of our shadow, we prevent re-acting, and we become observers who act with from the heart, not from our guts.

A healer is responsible and honest, and is inclined to service. A healer is generous and can, if necessary, and without detriment to herself, postpone (without eliminating) addressing her own needs to attend the needs of others.

To support physical healing processes of self and others, reiki practitioners need to understand the human body, illness and health from a different perspective. These are some of the ways in which the practitioner can support others:

1. Using the *laying on of hands* to promote changes in the energetic field of the person; this will manifest in different body dimensions.
2. Promoting changes in lifestyle, including proper nutrition, enough physical activity and stress reduction. These changes will be the expression of emotional and mental maturation that will, in its time, stimulate the development of other aspects of the self. When people start loving their bodies, they stop splitting soma and psyche.
3. Facilitating the identification of limiting patterns of behavior, beliefs and attitudes that are the source of sorrow, to raise the awareness that leads to the necessary changes that eventually will put the habit of suffering to an end.
4. Introducing the ideas of a multidimensional body, of the symptom as a symbol and of the frictions that occur between the different dimensions of the human being.
5. Inviting to look at illness as an opportunity to know ourselves better, understand our unresolved businesses and how the body manifests conflict. This is to help explore the secondary gains of illness.

To facilitate this task, the healer:
1. Provides information that the receiver is ready to integrate, according to his stage of development and level of consciousness.
2. Facilitates the access to knowledge that is already in the person. There are no such things as experts and dummies. We are experts

only on ourselves and our bodies. We possess a knowledge of which we sometimes are unaware and that we have perhaps never used. The healer is an educator[49] who facilitates the person's access to his own knowledge that he accumulated through life.

3. Stimulates the search for a conscious life.
4. Becomes a vehicle for integration and witnesses the transformation that takes place in the other.
5. Serves as a mirror so that the other can have an accurate image of self.
6. Uses her hands to transfer energy. The *laying on of the hands* is a catalyst for the transformation processes.

The healer as apprentice

I totally agree with Doctor Jorge Carvajal, when he says that a healer needs to be a constant and humble apprentice who cultivates detachment. In every single act of healing, my patient is my master and the act of healing, my learning experience. We must understand that each person is unique in his own circumstances and therefore we cannot classify the sufferer according to the suffering and prescribe treatments following a pre-established protocol.

Most prescriptions are just that, a pre-determined protocol that is used without taking into account the particular needs of a person or the circumstances that surround them. By maintaining the attitude of the apprentice, we leave preconceptions and book knowledge behind, to avoid error. Scientific research has demonstrated that we cannot separate the observer from that which is observed.

When we facilitate healing, the best path is to trust our intuition, take into account the information that the person provides us, and carefully assess the body and its circumstances. This way, we will use both our brain and our sensitivity. This requires being totally present when we are healing.

Let's talk about *presence*. First of all we need to be present in our body, aware of our posture, breathing, and surroundings. It takes a few minutes. Once we are in tune, we pay attention to the flow of our thoughts and the emotions that are connected with them, without judging

---

[49] Education, from Latin *educere*, means to pull out (what is inside).

or fighting them. Once we achieve mental presence, we become good observers and listeners, which allow our souls to freely flow. Then, our intuition will guide our doing.

We know that we are really present when we experience a state of consciousness that is similar to a meditative state. We feel one with our body and with all that surrounds us in the present moment. Nothing else exists; we are not worried or anxious or time conscious. We are totally aware of what the moment brings us and we absorb each "particle" of experience with fruition but without effort. It is a state of free flow, maximum alertness, and ease.

We do learn a hand position protocol in reiki. But it has to be considered a point of departure for the initiate. By *laying on the hands* often on self and others, the practitioner learns to feel the energy and gains enough confidence on her role as an instrument to channel the energy, and will herself to be guided by intuition.

Often the therapists, especially those who work with energy, talk about the importance of intention. Some propose that energy follows the mind and we should concentrate the mind on an intention to achieve a certain result. In my experience this belief might be problematic.

Who knows what the exact needs of the multidimensional body are? And if we say that the energy follows the mind, is then our thought what determines the changes? I think this might be a simplistic interpretation of the *law of attraction*.

I believe there is a big difference between the use of the faculties of the mind, which are at play when I am thinking, and the faculties of the spirit. When we apply reiki we are connected through the soul to the spirit – a *force* that links us all. Thus, it is not our thoughts that lead to the transformation of matter but rather it is my knowing (my certainty, my faith), which dwells in my soul when it's connected to the spirit, to which the world conforms itself.

The power of thought is limited; the power of the universal energy or spirit is limitless.

The therapeutic alliance

When we interact with the person receiving the treatment, we create what in psychology is known as a *therapeutic alliance*, which comprises implicitly or explicitly, a contract that defines our relationship. The terms of this contract include time frames, the role each of us will play, confidentiality issues, mutual respect, financial arrangements, if any, etc.

In each therapeutic alliance a power differential exists. The person who comes to us trusts us because he believes that we can help. They consider us *experts*. There is certain vulnerability in the person who accepts this dependent role and it can elicit a process called *transference*. It usually happens to patients when a parent-child relationship is unconsciously re-established and the client transfers to the therapist feelings and unresolved needs that do not belong to the present moment.

Transference is present in most relationships where there is a real or perceived power differential (boss, teacher, therapist).

In individuals who are unaware of or not psychologically able to handle these feelings, transference may become the dominant reality, causing frequent disappointment and rejection in many relationships, often followed by anger or withdrawal.

Countertransference is transference occurring in the opposite direction. It can adversely affect the therapeutic relationship and can be harmful to the client because it interferes with the understanding of the client's actual needs.

Sometimes *countertransference* reflects the unresolved infantile needs of the therapist and other times it refers to attitudes and feelings that are an unconscious response to what the client is transferring to the therapist. *Countertransference* can be used as a source of valuable information about either our unresolved issues or the needs of the client.

Countertransference might be occurring if a practitioner experiences:

√ More negative or positive emotional charge than usual towards a client.
√ Irritability or anger because the client is not getting better.
√ Thinking that your work is better/worse that others'.
√ Being exhilarated, depressed, tired, uneasy with a particular client.
√ Consistently feeling attracted to clients.
√ Feeling like doing favors to clients, outside the treatment context.
√ Having expectations of receiving praise.

The therapist needs to be alert to signs of transference and countertransference, learn from them, process unresolved needs and learn to detach from the outcome of the treatment.

Detachment and humility go hand in hand. We need to realize that ours is not the role of a savior. Sometimes all that is expected from us, all that we can do is respect the other person's process (something that maybe other people in their lives are not doing).

Healers should learn to be humble, renounce notoriety and protagonism. In a competitive society where we fall on the fallen because we cannot tolerate losers, we end up looking for celebrity to survive the pressure. We are motivated many times by a definition of success that is measured by academic achievements, titles or material possessions and thus it is difficult to become humble or to display humbleness.

Being humble implies that we understand that in the act of healing, we don't have a protagonist role. The healing is not about us but about the other and we will explore where his consciousness is to decide how to support him better. Even though we operate simultaneously in all the dimensions of our being, our consciousness tends to dwell in one of them. Usui said it in a different way. He believed that the mind and the spirit should be healed first and then the healing of the body would ensue.

When we perceive where the consciousness of the client resides, we can work with them from their own perspective. In a therapeutic relationship, the perception and the quality of the relationship, and trusting the proposed course of action sum up to become the key elements that impel the transformation and acceptance needed for the healing to take place.

If the person we are working with shows a constant preoccupation for the body, its functioning, its symptoms, medication, and nutrition it is most likely that their consciousness is in the physical level.

If a person speaks about his mood, interpersonal relationships, dissatisfaction with his job and expresses himself in terms of his feelings, this person most likely has his consciousness in the emotional realm.

Other people will show a preoccupation for intellectual issues, the economy, the causes of illness, different alternative treatments or science, suggesting that their consciousness resides in the mental body. The practitioner might even notice that this person has trouble being in touch with his feelings and will talk only about his thoughts.

And even others will have their consciousness on the spiritual body and will probably talk about God's will, displaying a humble disposition. They will request support for their inner process when feeling exhausted or need to clarify certain issues in order to advance in difficult times.

On some occasions we may examine, with the person we are treating, some aspects related to the degree of evolution they have achieved and it is then important to see how much responsibility they are assuming, which reflects the development of their chakras.

During treatments, the intelligence of the universal energy awakens the intelligence in the multidimensional body, which has been drowsy because of lack of stimulus or stress.

A healer seeks a rhythmic life

A healer learns to take care of herself, maintains a rhythmic life, a healthy nutrition, enough physical activity and makes sure that there is room in her life for activities that counteract stress. Because reiki is utilized most of all to support the self's healing process, after the initiation to the first degree, the master insists in the importance of physical care.

Under pressures and demands deriving primarily from cultural influences in modern life, we learn to segment or break up our natural way of moving and being. We develop protecting patterns of clenching. Understanding why and how we segment our movement helps us overcome restrictions that are the source of pain and discomfort.

I found it useful to give and receive Trager sessions and to learn moving with my whole body. The Trager approach is named after Dr. Milton Trager who developed and refined his method of movement re-education and rehabilitation for over 60 years. At age 18, he intuitively discovered certain principles that he later applied to his work with patients who had exhausted the possibilities of conventional medicine. After the treatment, many of these patients experienced increased mobility and freedom from chronic pain.

In a typical Trager session, clients lie on a massage table fully or partially clothed. The practitioner enters into a state of awareness similar to that of meditation (Milton Trager called it "hook up") and uses touch and movement to sense the client's tissues helping his body seek a freer, easier and more functional way of being and moving.

Mentastics, a term coined by Trager to refer to a self-care movement process of *mental gymnastics*, is a complementary way of exploring movement possibilities that the practitioner teaches the client. If practiced mindfully and regularly, mentastics helps to redefine body image, which increases feelings of confidence, self-awareness and functional ability.

After I became certified in the Trager approach, I learned to bring focus to the way I move, creating awareness not only of the restrictive patterns that I have developed over time due to habitual posture and traumas, but also of newer possibilities.

I know that whenever I move, the tension that my back and hips accumulate while I sit in front of the computer dissipates. I usually give myself a break every 45 minutes to change posture and walk around.

The importance of pausing was also introduced to me by the Trager instructors. Pauses during the Trager session allow the nervous system to integrate the information that movement and touch provided. On a larger scale, pausing refers to our need to rest. It's during sleep that the body regenerates and repairs tissues. It's during rest that the hippocampus sends to the cerebral cortex the information that will become our permanent memory.

A rhythmic life implies shifting between activity and repose. Pauses allow us to integrate what we have just learned.

Life is made up of cycles that we must learn to respect. These cycles are present in all aspects of life, seasons, day and night, the different stages we go through in a lifetime, our *chronobiology* (our biological day cycles).

In the beautiful vignette *Watermills Village* (In Dreams), by Japanese director Akira Kurosawa, a healthy 93 year-old villager who seems to be cleaning the blades of a watermill wheel, answers to a visitor who inquires if there is electricity in the village.

"No need for it," the old man says. "People get too used to convenience. They think convenience is better. They throw out what is really good."

"But what about lights?" the visitor asks.

"We've got candles and linseed oil."

"But night is so dark!" the visitor says.

"That's how it's supposed to be. Why should we make it as bright as the day? I wouldn't like nights so bright that you couldn't see the stars," the old man says.

Modern life has us so used to convenience! And with convenience comes a state of agitation. We even feel certain remorse when we enjoy a conversation after lunch or take a siesta. We no longer work for a living, we live to work! And we forget to listen to the body and respond to its demands. We take meals and go to sleep by the clock instead of responding to the need to eat or the need to rest.

And last but not least, the healer respects life in all its forms and expressions. This is coherent with her work as a healer that in its essence is to promote life.

Primum, nil nocere

My best definition of moral conduct is summarized in one single imperative that Hippocrates wanted his disciples to follow: *Primum, nil nocere*, first do no harm. Several clauses in the Hippocratic Oath refer to this same question. See below.

Translated to the approach and the subject matter of this book, a reiki practitioner should learn to respect the *inner healer*. I think that our contribution is in part to inform and to encourage the patients to be informed about their bodies so that they will be cautious before accepting invasive options that involve putting chemicals in the body or undergoing surgery when not strictly necessary.

Health is propped up by three basic pillars: adequate nutrition, an active life and stress reduction. Therefore, the task of the healer or therapist is focused on bolstering these pillars.

A healer does not diagnose but rather maintains a therapeutic attitude at all times. In cases where clients come with a diagnosis, the healer helps the client avoid his identification with labels.

A *diabetic* is no one else but someone who is having trouble maintaining blood sugar in normal levels. The idea of being a diabetic evokes a condition that can never be resolved. But the idea that our bodies have gone out of whack because of our lifestyle may be an invitation to take care of our health and plants the seed of hope.

If we can modify our habits and the unfavorable conditions existing in our lives, we can improve our health. I have seen on many occasions how this approach renders positive results, especially when the patient is treated by a person he can trust.

Hippocratic Oath – Classical version

I swear by Apollo Physician and Asclepius and Hygeia and Panacea and all the gods and goddesses, making them my witnesses, that I will fulfill according to my ability and judgment this oath and this covenant:

To hold him who has taught me this art as equal to my parents and to live my life in partnership with him, and if he is in need of money to give him a share of mine, and to regard his offspring as equal to my brothers in male lineage and to teach them this art - if they desire to learn it - without fee and covenant; to give a share of precepts and oral instruction and all the other learning to my sons and to the sons of him who has instructed me and to pupils who have signed the covenant and have taken an oath according to the medical law, but no one else.

I will apply dietetic measures for the benefit of the sick according to my ability and judgment; I will keep them from harm and injustice.

I will neither give a deadly drug to anybody who asked for it, nor will I make a suggestion to this effect. Similarly I will not give to a woman an abortive remedy. In purity and holiness I will guard my life and my art.

I will not use the knife, not even on sufferers from stone, but will withdraw in favor of such men as are engaged in this work.

Whatever houses I may visit, I will come for the benefit of the sick, remaining free of all intentional injustice, of all mischief and in particular of sexual relations with both female and male persons, be they free or slaves.

What I may see or hear in the course of the treatment or even outside of the treatment in regard to the life of men, which on no account one must spread abroad, I will keep to myself, holding such things shameful to be spoken about.

If I fulfill this oath and do not violate it, may it be granted to me to enjoy life and art, being honored with fame among all men for all time to come; if I transgress it and swear falsely, may the opposite of all this be my lot.

Let's remember that we can program information in our neocortex, pre-determining bodily responses. When a person identifies with a diagnosis, it will become the core of his life and he will construct his reality around it. It will even become an excuse to avoid situations and not look for solutions.

When we are facilitating a client's process by suggesting that the symptom might be a symbol, we help him focus on healing, not only on curing, according to the definitions that we have already discussed.

Illnesses, physical or mental, create secondary psychosocial gains of which we are not aware. We can support a person's discovery of healthy expressive channels so that the person's body doesn't *have to* convert a conflict into disease.

From a dynamic vision of the multidimensional body I am convinced that the body is always in search for balance and that balance depends on the capacity of the organs to communicate among them.

Communication is also pervasive among our different bodies. On the energy level (subtle bodies), we could say that we are embedded in energy and that this energy delineates blueprint maps that the body uses as a reference.

At the physical level, we know that interaction between genes determines and regulates the production of certain messenger molecules and that these molecules travel around our body in search for matches, to maintain stable conditions and grant survival.

This vision supports the idea of an intelligent body that learns and evolves throughout a lifetime, using inherited resources, and has the capacity for self-regulation, self-reparation and self-regeneration. Everything is interconnected; we belong to the Earth and the Earth to the Universe and our body and planet deserve all respect and consideration.

We have a soul that connects with an infinite source of wisdom.

But even if we have the best of intentions to respect our bodies, we do not exist in isolation. We are part of a community with certain characteristics and as it is today, these characteristics have become obstacles in our path to health and cannot be defeated individually.

With a better understanding of how emotions and inner harmony determine health or illness, we can take responsibility about our ways of relating with ourselves, others and the cosmos.

Annexures

Annexure 1 - Research

Among the obstacles encountered by researchers are the lack appropriate equipment to measure reiki energy and the lack of funding. Notwithstanding, research has demonstrated reiki's beneficial effects. The *World Health Organization* recognizes reiki as one of the alternative practices that has an effect on the *biofield*. In many hospitals of the world, reiki is offered to cancer and terminal patients.

Research seems to have focused on two important areas. One refers to the clinical and physiological effects of reiki. Researchers have tried to quantify the measurable changes in either the biological activity of the subject or the course of a specific condition. The other area is the study of the healing phenomena.

Biochemical and physiological effects of reiki and other energy healing modalities have been reported. English zoologist and healer Tony Bunnell has published studies about the changes in enzymatic activity at the cellular level when a hand-on healing treatment is applied. Doctor Daniel Benor made an extensive review of existing publications about the efficacy of hands-on healing and found that at least two thirds of the publications reported significant positive results. The studies included both direct *laying on of the hands* and distance healing. The subjects of the study ranged from microorganisms, to plants and animals to humans.

Reiki Master William Lee Rand has published in his Web site (www.reiki.org) several abstracts about studies that demonstrate the healing properties of reiki. For example:

Nurse Wendy Wetzel found that hemoglobin levels increased in people who were initiated to reiki. These results have been replicated in other studies.

Janet Quinn, from *South Carolina University* found that levels of anxiety lowered after the *laying on of the hands*.

Daniel Wirth, from *Healing Sciences International* in California, demonstrated on students with minimal self-inflicted wounds that reiki contributed to accelerate the process of wound healing.

The following are some of the effects of reiki already supported by research findings:

√ It alleviates pain and anxiety
√ Reduces stress and fatigue

√ Can stimulate the immune system
√ Can stimulate the production of red blood cells in people with low counts
√ Accelerates wound healing

Several authors explain the healing phenomenon based on a possible synchronicity that has been observed between the magnetic waves of the Earth, the brain waves and the waves emitted by the healer's hands. Studies with athletes and musicians have shown that they achieve their best performances when they enter a state known as *the zone*, which has been depicted as a moment of total inner wholeness where the person is connected with the whole. In *the zone* the person's performance becomes almost effortless. This state is similar to the consciousness achieved by people who meditate and is also frequently found among healers.

The most common experience for those who receive reiki is relaxation. Both arousal and relaxation states involve the whole body but their effects are more evident in the autonomic nervous system and the endocrine and musculoskeletal systems. It seems that when the muscle tension decreases, this relaxes the mind and promotes the production of endorphins, the body's natural sedatives.

Several explanations have been proposed for the relaxing effect caused by reiki. One is that the person who is receiving reiki is pausing in her activities and focusing on the experience of reiki. He is being touched by someone who cares, and it feels good. There is plenty of research about the effect of caressing the skin and how the strokes activate the nerve receptors in the skin, inducing a relaxation response that is mediated by the parasympathetic system. Every pleasant situation makes our body produce endorphins.

Another factor that seems to influence the effect of reiki is the quality of the relationship with the reiki practitioner. Unconsciously, we are always evaluating the quality of a relationship. When we first enter in a relationship the body responds with some tension. When we start to trust the other, we begin to relax.

Doctor Herbert Benson, founder of the *Mind-Body Medical Institute,* has investigated the relaxation response in the human body. Initially, his studies were focused on the effects of transcendental meditation that he found beneficial for the heart and then, when he found that people would keep practicing meditation if prayer was added, he investigated the therapeutic value of faith.

There are also some hypotheses that seem metaphysical and cannot presently be supported with scientific research.

Reiki practitioners find reiki useful to support, without replacing, any medical treatment for acute or chronic illnesses and it offers relief to terminal patients.

Since 1994 I have practiced reiki on myself almost daily and for all sorts of situations and ailments. I have not taken any pharmaceuticals, not even an aspirin or an antibiotic in these years. I have occasionally used vitamins and other nutritional supplements in times of extreme stress or when my body was signaling a deficiency. Nowadays food is produced in a way that depletes the soil of nutrients and what we eat is often times poor in nutritional value, even if our diet is vegetarian, as mine is.

Most of the evidence to support the benefits of reiki is anecdotal. I've heard the *not enough evidence* argument once and again in medical lectures where physicians are asked about recently published research that leads to advising in favor or against of the use of certain medications, procedures, or foods. I've heard them say there is *not enough evidence* to using or stop using them. This usually means that the results don't come from what they would deem reliable sources, are not the product of controlled studies, or that the evidence is mostly anecdotal (observational evidence). Insufficient evidence prevents physicians from either adopting certain treatments or recommendations that could be beneficial, or from discontinuing prescriptions of certain medications. And this is called best practice.

Presently, most physicians rely mostly on information coming from their profession's approved sources, usually disregarding independent investigators and studies published abroad. Unfortunately, many doctors seem more inclined to advocate for the use of medication than to encourage alternative therapies and lifestyle changes. I really hope that the time will soon come in which the thousands of anecdotes collected by alternative practitioners will sum up to the critical mass necessary to introduce a different mentality about how to safely regain the body's wisdom.

Annexure 2 - Nutrition

What follows are the American Cancer Society's guidelines devised for cancer prevention and treatment. They include nutritional as well as physical activity recommendations. These are endorsed by other organizations such as the American Heart Association and the American Diabetes Association. Their independent research has come to similar conclusions: that optimal functioning of the body largely depends on a proper nutrition.

Grains, vegetables and fruits provide the body with the most antioxidants, needed to counteract the damaging and aging effect of free radicals. As we said elsewhere, nutrition is a basic pillar of health.

Visiting www.mypyramid.gov you can find the most recent guidelines published by the *U.S. Department of Agriculture*. Clicking on *My Pyramid Plan* you can get a personalized recommendation on the amount of each food group you need daily. Just enter your age and gender (unfortunately it doesn't take into account your size, which would provide a more accurate guide). *The National Guideline Clearinghouse* also summarizes nutritional guidelines for heart disease, cancer and diabetes. Visit: www.guideline.gov

Recommendations for individual choices

1. Eat a variety of healthful foods with an emphasis on plant sources.
2. Eat five or more servings of a variety of vegetables and fruits each day.
    √ Include vegetables and fruits at every meal and for snacks.
    √ Eat a variety of vegetables and fruits.
    √ Limit French fries, snack chips, and other fried vegetable products.
    √ Choose 100 percent juice if you drink fruit or vegetable juices.

3. Choose whole grains in preference to processed (refined) grains and sugars.
    √ Choose whole grain rice, bread, pasta, and cereals.

√ Limit consumption of refined carbohydrates, including pastries, sweetened cereals, soft drinks, and sugars.

4. Limit consumption of red meats, especially those high in fat and processed.
   √ Choose fish, poultry, or beans as an alternative to beef, pork, or lamb.
   √ When you eat meat, select lean cuts and have smaller portions.
   √ Prepare meat by baking, broiling, or poaching rather than by frying or charbroiling.

5. Choose foods that help maintain a healthful weight.
   √ When you eat away from home, choose foods that are low in fat, calories, and sugar, and avoid large portion sizes.
   √ Eat smaller portions of high-calorie foods. Be aware that "low-fat" or "nonfat" does not mean "low-calorie," and that those low-fat cakes, cookies, and similar foods are often high in calories.
   √ Substitute vegetables, fruits, and other low-calorie foods for calorie-dense foods such as French fries, cheeseburgers, pizza, ice cream, doughnuts, and other sweets.

6. Adopt a physically active lifestyle.
   √ Adults: engage in at least moderate activity for 30 minutes or more on five or more days of the week; 45 minutes or more of moderate-to-vigorous activity on five or more days per week may further enhance reductions in the risk of breast and colon cancer.
   √ Children and adolescents: engage in at least 60 minutes per day of moderate-to-vigorous physical activity for at least five days per week.

7. Maintain a healthful weight throughout life.

   √ Balance caloric intake with physical activity.
   √ Lose weight if currently overweight or obese.

8. If you drink alcoholic beverages, limit consumption. Men, no more than two drinks per day and women, no more than one drink per day.

Good nutrition should provide plenty of antioxidants to counteract the deleterious effect of free radicals in the body. Veggies and fruits are the main source of antioxidants.

Anti-inflammatory and pro-inflammatory foods

Twenty years ago, a doctor developed rheumatoid arthritis, which limited his performance as a surgeon. In search of a cure, and after several months of taking anti-inflammatory medication that badly upset his stomach, he opted for alternative treatments.

His new diet excluded meat and dairy, even though his former primary doctor recommended against it. He also tried meditation, massage, acupuncture and homeopathy. And when after a few months his condition started to really improve, instead of going back to work as a surgeon, he decided to train in the same treatments that helped him. Illness also gave him the opportunity to learn that he needed to change his levels of stress in order to maintain health.

For many years, part of popular wisdom was that meat and dairy have a negative effect on arthritic conditions. But recently, mainstream medicine is corroborating the relationship between certain foods and the exaggerated inflammatory response seen in conditions such as arthritis.

We tend to see inflammation as an unwanted reaction from the body. Notwithstanding, inflammation is the normal bodily response to harmful stimuli. Without inflammation, the body would not be able to repair damaged tissues or respond to an invader.

Scientists are excited about recent discoveries about what we could call pro-inflammatory and anti-inflammatory foods. In its February 2004 edition, for example, the *Tufts University Health & Nutrition Letter* published the article *Anti-Inflammatory Eating,* reporting on studies that correlate three medical conditions – high blood pressure, coronary disease and arthritis – with nutritional habits.

The fats and oils that we use in our food are prostaglandin precursors in the body. Some of these substances hold back the inflammatory response (the ones coming from omega-3 fatty acids,

192

present in olive and flax oil and salmon) while others exacerbate the inflammatory response (like those that come from omega-6 fatty acids present in animal fat, corn, sunflower and cotton oils). Many processed foods are prepared with fats rich in omega-6 fatty acids.

White rice, white bread, refined sugar and other foods that tend to produce acid in the body, such as meat and dairies, also elicit an inflammatory response. On the other hand, most fruits and vegetables, legumes, tofu and seeds help to keep it at bay.

Fruits like papaya and pineapple rich in certain enzymes (papain, bromelain) contribute to curb inflammation. Ginger and red tart cherries are in this category as well.

Bottom line is science is arriving at the conclusion that a balanced diet rich in anti-inflammatory foods and limited in pro-inflammatory foods could contribute to the improvement of conditions such as arthritis and cardiovascular disease.

We should not eat ourselves to death

With the introduction of computers, videogames, remote controls and 24-hour TV programming, we are moving less and less. Lack of physical activity is a major cause of weight gain.

In 2003, the federal government's *Center for Disease Control* (CDC) rang the alarm bell about America's obesity epidemic. Headlines such as "How to drop 40 pounds in a week," "Shave off waist inches easily" or "Walk your way to a better burn" invite people to lose weight effortlessly have proliferated in magazines and health newsletters. Others such as "Studies link obesity with prostate cancer" or "Abundant evidence links overweight and obesity with impaired health" warn about the danger of being overweight or obese.

In 2000, individuals who were overweight or obese spent more than $35 billion a year on weight loss products and services. That figure – and the nation's collective waistline – has expanded since then. Americans are trying to lose weight or prevent weight gain by buying videos, books, dietary supplements or any other product that promises miraculous results.

According to the CDC, physical inactivity and nutritional deficiencies are the cause of the obesity epidemic, and only lifestyle changes can guarantee healthier lives and leaner bodies.

Lifestyle changes required to keep a desirable weight and prevent the diseases linked to obesity such as heart ailments and diabetes, include eating fewer calories, consuming less processed food, being more physically active and reducing stress levels. But individuals alone cannot shoulder all the responsibility.

There are chemicals in food, such as monosodium glutamate, that have been related to weight gain. Another problematic product is fructose. German researchers found a relationship between consuming *high fructose corn syrup* and weight gain. Fructose does not need insulin to enter cells and is easily stored as fat. Many processed foods and most sugary drinks contain corn syrup.

In the past five years, the food industry has used advertising to sell products based on their claimed potential to contribute to weight control or weight loss. For example, the *National Dairy Council* spent $200 million promoting the idea that milk helped to reduce weight. In June 2005, the *Physicians Committee for Responsible Medicine* filed suit, contending the industry's advertisement was deceptive. The weight-loss campaign was based on studies conducted by Michael B. Zemel, a professor at the University of Tennessee. Who had funded his research? The dairy industry.

In a 2002 report on the current trends in weight-loss advertising, a *Federal Trade Commission* staff group, with the assistance of the *Partnership for Healthy Weight Management* (consisting of experts from the scientific community, academia, health care, government, commercial enterprises and other organizations), examined false and misleading claims in the advertising of weight loss products and services. They found that nearly 40 percent of the ads in their sample "made at least one representation that almost certainly is false."

We have built a society that looks for easy, effortless solutions to difficult and complicated matters. But in our search for comfort we have paradoxically increased the risks to our health.

Lite might not be the best for your health

Sodas, beers, precooked meals and even genetically engineered avocados: all of them have gone "light" in the past few years. And, what does *lite* really mean? By definition, when we talk about *lite* we're referring to a product that contains at least 50 percent less fat or 50 percent fewer calories furnished by the fat. *Lite* is also a word used for drinks with lower levels of alcohol or for meals with lower levels of salt.

But, is *lite* necessarily healthier?

We can avoid some mistakes when we're trying to follow healthy recommendations.

Take for example *lite* salad dressings. When trying to stay away from the 120 calories and 11 grams of fat per spoon in mayonnaise or the 90 calories and 18 grams of fat in a serving of blue cheese dressing, people switch to light versions of the products, although seldom reading the nutrition facts on the label. Should they read the label, they would realize that most of these *lite* dressings are lower in calories indeed, but heavy in sugar and salt and actually only light on nutritional value.

A better choice would be a simple oil-and-vinegar dressing that, although high in calories contains lots of heart-healthy mono-unsaturated fatty acids and no saturated fat. My own recipe, sour cream and ketchup with a few drops of olive oil, has less than 50 calories per teaspoon, contains a nice balance of omega-3 and omega-6 fatty acids and tastes good.

Another example is found in meals labeled "low-sodium." Processed food is the main source of excess salt in our diet, which has been associated with cardiovascular disease. The law requires that the manufacturer cut only 25 percent of the sodium from the original product. Solely the products marketed as "low-sodium" have the recommended less than 140 milligrams of sodium per serving.

People on low-calorie diets who love sodas have turned to diet sodas containing aspartame instead of sugar, but there are 92 documented symptoms related to aspartame, from headaches to death. Birth defects, lupus and multiple sclerosis-like symptoms have been linked to aspartame poisoning, although studies offer contradictory results.

Sharon Fowler and colleagues at the *University of Texas* in San Antonio reported earlier this year that people who drink diet soft drinks not only don't lose, but gain weight. The team reviewed eight years of data on 1,550 Mexican-American and non-Hispanic white Americans aged 25 to 64. Of the 622 study participants who were of normal weight at the beginning of the study, about a third became overweight or obese.

"What didn't surprise us was that total soft drink use was linked to overweight and obesity," Fowler told WebMD. "What was surprising was when we looked at people only drinking diet soft drinks, their risk of obesity was even higher."

Interestingly enough, when the researchers analyzed their data, they found that nearly all the obesity risk from soft drinks came from diet sodas.

"There was a 41 percent increase in risk of being overweight for every can or bottle of diet soft drink a person consumes each day," said Fowler.

The problem is that you cannot fool the body. Food with low nutritional value, compared to fresh food, won't keep you satisfied for long. Researchers hypothesized that diet sodas stimulate appetite.

It is important to become wise buyers, not only making intelligent choices at the supermarket, but also not "buying" (believing) everything you hear or read. Develop your own criteria, do some research to inform yourself and then create healthy habits including buying reliable, fresh or low-processed products.

Why keep changing brands and improvising meals with differing calorie and fat values because of the commercials you saw on TV? Every so often, people trying to shed just a few pounds became so self-conscious of their appearance and obsessed over food calories and fat content that they ended up battling eating disorders and obesity. The best path to health is to find what works for you and stick to your own healthy choices.

If stricter regulations prevented the food industry from selling foods and beverages filled with sugars, trans fats, saturated fat, salt and chemical additives, those companies would suffer billion dollar losses. Their profits are built on promoting sickening habits. Some of them try to compensate by educating the public about eating healthy!

I wonder if the concept of freedom actually encompasses allowing the few to make profit from the many by introducing products that are not beneficial to their health.

References and Bibliography

Arjava F., Reiki, the Legacy of Dr. Usui. Lotus Light, 1999.

Benor, DJ, Healing Research, Volume I, Scientific Validations of a Healing Revolution, Southfield, MI. Vision Publications, 2001.

Bloom, J. Junk Science as Much a Part of "Fat epidemic" as junk food. In: Advertising Age. 76, (Jan, 24, 2005), p25.

Bodynamic Institute USA, About Bodynamic analysis. In www.bodynamicusa.com/AboutBDYN.html. 1999.

Book, H., Brief Psychodynamic Psychotherapy, American Psychological Association, Washington, 1997.

Brennan, B. A. Manos que curan, Latinoamericana, Colombia, 1990.

_____ Hágase la Luz, Martínez Roca, Barcelona, 1994.

Brines R. Neuroendocrineimmunology today. Inmunology Today, 1994.

Buhler, R., Pain and Pretending. Thomas Nelson Publishers, Nashvill, 1988.

Caldwell, C. Getting our Bodies Back, Shambahala, Boston, MA, 1996.

Capra, F., The Immune System our Second Brain. In http://resurgence.gn.apc.org/articles/capra.htm

Carter, R. El Nuevo Mapa del Cerebro. Integral, Italia, 1998.

Carvajal, J. Un Arte de Curar. Norma, Bogotá, 1995.

Castaneda, C. El Fuego Interno. Edivision, México, 1998.

Chang J., Fisch J., Popp F., Biophotons. Book News, Inc., Portland, OR, 2002.

Chopra, D. Ageless Body, Timeless Mind: The Quantum Alternative to Growing Old. Three Rivers Press, NY, 1993.

Cohen, P. You are what your mother ate. In: New Scientist, 179, 8/9/2003, p14.

Cousins N., Anatomy of an Illness, Bantam Books, NY, 1979.

Dethlefsen, T. & Dahlke, R. The Healing Power of Illness: The Meaning of Symptoms and how to Interpret Them. Vega Books, 2002

DSM IV, American Psychiatric Association, Washington, 1994.

Freud, S. La Histeria. Alianza editorial. 3a. ed. Madrid, 1970.

Fritz, S. Mosby's Fundamentals of Therapeutic Massage, Mosby, Saint Louis, 2000.

Gerber R. La curación energética. Intermedio Editores/Robin Book, Bogotá, 1993.

Gleick, J. Chaos: Making a new science. Penguin, NY, 1987.

Khan A, Warner H.A, and Brown W.A. Symptom Reduction and Suicide Risk in Patients Treated With Placebo in Antidepressant Clinical Trials. An Analysis of the Food and Drug Administration Database. *Arch Gen Psychiatry.* 2000;57:311-317.

Juhan, D., Job's Body, A Handbook for Bodyworkers. Station Hill Press. NY, 1987.

Kouguell, Mind/Body Therapy en Magazine for Hypnosis and Hypnotherapy. In www.hypnos.co.uk/hynomag/kouguel15.htm.

Levine, P.A. Waking the Tiger, Healing Trauma, North Atlantic Books, Berkeley, CA. 1997.

Locke, S., Colligan, D. El médico interior. 3a. ed. Buenos Aires. Sudamericana, 1986.

Miss C. Anatomy of the Spirit.The Seven Stages of Power and Healing. Bantam Books, NY, 1997.

Moreno, J. L. Psicoterapia de grupo y psicodrama. Fondo de Cultura Económica, México, 1959.

Newberg, A., D'Aquili, E., Rause, V. Why God Won't Go Away. Science and the Biology of Belief, Ballantine Books, NY, 2001.

Oschman J.L., Oschman, N. How Healing Energy Works. www.bodywork–res.com

Osho, Zen, The Path of Paradox, St. Martin's Griffin. NY, 2001.

Page, J. CranioSacral Therapy 21 years on. Upledger Institute. In www.upledger.com/cstam.htm, 2000.

Pert, C.B. Molecules of emotion, the science behind Mind–Body Medicine, Touchstone edition, NY, 1999.

Quinn, D., Ishmael, Bantam, NY, 1995.

Sagan, C. Cosmos. Planeta. Barcelona, 1980.

_____. Los dragones del Edén. Grijalbo, NY, 1977.

Shea, M.J., Somatic Psychology for Bodyworkers, Shea Educational Group, 2000.

Sheperd P., Body–Mind Defences. In www.trans4mind.u–net.com/ transform4.13.htm., 2000.

Stanihurst, R. Toque de Alquimia, (texto del siglo XVI editado por Pedro Rojas García, «Azogue», n° 4, 2001. In   http://come.to/azogue).

Stein, H.F. What is Therapeutic in Clinical Relationships? In: Family Medicine, Vol. XVII, (Sept–Oct, 1985).

The Hippocratic Oath, Translation and Interpretation from the Greek by Ludwig Edelstein. Baltimore, Johns Hopkins Press, 1943.

Thibodeau, G.A., Patton, K. Anatomy & Physiology. Mosby, St Louis, 1999.

Villoldo, A. Los Cuatro Vientos. Planeta, Buenos Aires, 1992.

Vox, Diccionario General Ilustrado de la lengua española. Bibliograf, Barcelona, 1982.

Watzlawick, P., Beavin, J., & Jackson, D. Teoría de la comunicación humana. Herder, 9a. ed., Barcelona, 1993.

Weigent, D. A. Carr D. J. & Blalock J. E. Bidirectional communication between the neuroendocrine and immune systems. Common hormones and hormone receptors. In: Annals of the New York Academy of Sciences, Vol 579, Issue 1 17-27, 1990.

Web Sites researched:

On Buddhism

http://www.budsas.org/ebud/whatbudbeliev/78.

http://zencomp.com/greatwisdom/

On Physics and energy

http://www.chiexplorer.com

http://imagine.gsfc.nasa.gov/docs/science/know_l1/emspectrum.

http://www.kheper.net/topics/chakras/nadis.html

http://www.nal.usda.gov/fnic/etext/000062.html

On health and mind/body relationship

http://www.americanheart.org

http://www.analesdemedicina.com

http://www.bodywork–res.com

http://cancer.org

http://www.coxnews.com

http://www.diabetes.org

http://www.gsdl.com/news/connections/vol2/conn19981014.html

http://www.mbmi.org/research/default.asp

http://www.night–thunder.com/brainbal.html

http://www.robertogiraldo.com/esp/articulos

http://www.rethinkingaids.com

http://repairfaq.cis.upenn.edu/sam/icets/basicp.htm

http://www.the-scientist.com/news/20040618/01

On Nutrition

http://www.aidsnutrition.org/

http://business.fortunecity.com/mcca w/204/id36.htm

http://www.gmhc.org/health/nutrition.html

http://www.thebody.com/cdc/faqnut.html

On reiki and subtle bodies

http://angelReiki.nu/ryoho/shoden.html

http://angelReiki.nu/ryoho/tendai.htm

http://www.asunam.com/index.html

http://www.lifepositive.com/Body/energy–healing/Reiki/Reiki–
    energy.asp

http://www.oshogulaab.com/OSHO/TEXTOS/LIBROCHAKRAS.html

http://Reikihistory.topcities.com/Usui.html

http://www.soulworkings.com/Reiki_research.html

http://www.usuiReiki.fsnet.co.uk/myhistory.html

http://www.geocities.com/drukmar/

Index

acid-base equilibrium, 106
acidity, 25, 105, 130
acupuncture, 7, 34, 49, 86, 87, 98, 99, 134, 136, 192
adapt, 15, 50, 61, 62, 90, 127, 129, 138, 144
adaptive challenges, 17
addiction, 4, 5, 54, 135
AIDS, 8, 121
allopathic, 12
Allopathic medicine, 25
alternative therapies, 8, 189
amygdala, 124
Antibiotics, 25
antidepressants, 37, 114
anti-inflammatory, 8, 28, 70, 94, 120, 130, 131, 133, 192, 193
arthritis, 14, 34, 120, 128, 130, 147, 148, 192, 193
attachments, 79, 156
autoimmune, 120, 128, 130
awareness, 7, 14, 22, 24, 25, 26, 27, 45, 82, 83, 94, 100, 111, 126, 127, 140, 152, 160, 161, 166, 171, 172, 174, 180
Ayurveda, 67, 79, 108
Bach flower essences, 5
balance, 1, 2, 8, 14, 15, 18, 22, 29, 34, 36, 37, 50, 52, 59, 60, 61, 67, 70, 80, 87, 91, 94, 100, 105, 109, 114, 116, 118, 119, 126, 128, 131, 135, 137, 139, 144, 147, 149, 150, 151, 152, 183, 195
becoming aware, 37
biological medicine, 7, 141
biophotons, 74
body, 3, 1, 2, 3, 4, 5, 6, 7, 9, 12, 13, 14, 15, 16, 17, 18, 19, 20, 21, 22, 23, 24, 25, 26, 27, 28, 29, 30, 31, 32, 33, 34, 35, 36, 37, 38, 43, 50, 51, 52, 54, 55, 56, 57, 59, 60, 61, 62, 63, 65, 67, 68, 69, 70, 71, 72, 73, 74, 76, 77, 78, 79, 80, 81, 82, 83, 86, 87, 88, 89, 90, 91, 92, 93, 94, 95, 96, 97, 98, 99, 100, 101, 102, 103, 104, 105, 106, 107, 109, 110, 111, 112, 113, 115, 117, 118, 119, 120, 121, 122, 123, 124, 125, 126, 127, 128, 129, 130, 131, 132, 133, 134, 135, 136, 137, 138, 139, 140, 141, 142, 143, 144, 146, 147, 148, 150, 151, 152, 154, 156, 158, 159, 160, 161, 165, 166, 167, 170, 174, 175, 176, 178, 179, 180, 181, 182, 183, 188, 189, 190, 192, 193, 196, 200
brain waves, 76, 77, 188
Buddhism, 79, 80, 154, 157, 200
cabala, 79
cancer, 8, 20, 21, 34, 35, 56, 91, 93, 121, 128, 130, 147, 149, 187, 190, 191, 193, 200
cardiovascular disease, 34, 35, 70, 130, 193, 195
chakras, 67, 70, 79, 80, 81, 82, 83, 84, 85, 86, 87, 160, 167, 179, 200
change, 5, 13, 15, 19, 20, 24, 32, 36, 41, 45, 46, 47, 50, 54, 62, 63, 66, 76, 81, 86, 92, 102, 104, 118, 133, 134, 146, 152, 162, 163, 164, 167, 180, 192
Chaos Theory, 8, 145, 146
chemotherapy, 21
chiropractic, 49

cholesterol, 34, 35, 51
colitis, 32, 33, 120, 147
commitment, 13, 17, 46, 48, 93, 146, 152, 174
communication, 8, 17, 18, 30, 32, 33, 34, 43, 51, 56, 63, 65, 67, 68, 69, 70, 74, 85, 86, 89, 96, 98, 100, 101, 106, 112, 113, 119, 121, 123, 124, 132, 144, 152, 200
communication between organs, 32, 34, 43, 121, 144, 152
communication system, 18, 70, 86, 96, 98
communities, 21, 29, 38, 44
complementary medicines, 49
consciousness, 12, 14, 16, 17, 62, 75, 76, 77, 82, 85, 89, 115, 117, 118, 141, 143, 144, 153, 167, 171, 175, 176, 178, 179, 188
cortisol, 29, 115, 116, 124, 125, 127, 128, 130, 143
curing, 1, 12, 25, 91, 183
death, 34, 79, 137, 148, 153, 157, 193, 195
depression, 32, 37, 114, 115, 116, 126, 140, 144
detachment, 79, 175
development, 20, 21, 26, 43, 48, 64, 80, 81, 83, 85, 86, 92, 97, 123, 146, 157, 158, 164, 169, 171, 174, 175, 179
diabetic, 20, 23, 27, 181
diagnosis, 2, 14, 16, 23, 29, 30, 31, 50, 89, 132, 138, 139, 148, 181, 183
dialectic, 63, 92
diarrhea, 28, 33, 34, 61, 68, 89
diet, 2, 4, 5, 8, 15, 20, 24, 32, 33, 36, 38, 39, 40, 54, 55, 56, 60, 65, 67, 111, 130, 139, 189, 192, 193, 195, 196
digestion, 4, 33, 107, 108, 128
divine, 13, 17, 80, 81, 170

drinking water, 42, 44
dynamic, 59, 61, 63, 64, 65, 68, 74, 88, 94, 95, 114, 139, 147, 149, 164, 183
earth, 41, 67, 174
eating disorders, 35, 196
electromagnetic, 1, 51, 52, 55, 56, 71, 72, 73, 77, 138
emotions, 33, 41, 71, 78, 81, 84, 85, 90, 101, 105, 107, 117, 118, 123, 124, 128, 139, 142, 144, 148, 176, 183
endocrine system, 33, 109, 113, 115, 119, 120, 123
endorphins, 32, 41, 102, 116, 125, 135, 140, 188
energy, 1, 3, 4, 6, 12, 13, 15, 16, 17, 27, 28, 30, 34, 46, 51, 60, 65, 67, 69, 71, 72, 73, 76, 78, 79, 80, 82, 83, 85, 86, 87, 90, 92, 95, 98, 99, 100, 101, 103, 104, 107, 115, 127, 137, 139, 142, 143, 144, 145, 146, 147, 151, 156, 158, 160, 161, 162, 165, 166, 167, 169, 170, 175, 176, 177, 179, 183, 187, 199, 200, 201
enlightenment, 26, 78, 157
environment, 2, 17, 20, 21, 22, 26, 28, 46, 47, 50, 55, 60, 61, 65, 83, 86, 96, 97, 103, 104, 112, 126, 127, 136, 139, 141, 144, 162
Europe, 35, 66
Everglades, 44, 53
exercise, 18, 24, 55, 73, 102, 103, 116, 126, 173
fatty foods, 33
fear, 39, 43, 47, 52, 83, 93, 94, 100, 118, 138, 142, 148, 156, 158, 169
Feedback loops, 65
fiber, 32, 33, 55
Florida, 42, 44, 45, 46

free radicals, 50, 51, 55, 56, 190, 192
function, 1, 25, 34, 42, 50, 51, 61, 64, 65, 66, 70, 86, 88, 89, 92, 93, 97, 100, 101, 102, 105, 110, 113, 114, 115, 122, 123, 124, 129, 131, 135, 150, 151, 152
functional, 26, 93, 180
genes, 8, 18, 19, 20, 21, 92, 111, 112, 116, 183
genetics, 19, 21, 127
global warming, 40, 45, 46, 47, 52, 146
Global warming, 45
globalization, 30, 48, 59
greed, 40, 41, 42, 83
green house emissions, 45
guilt, 23, 27, 91
habits, 18, 19, 26, 29, 38, 40, 41, 106, 118, 126, 129, 130, 141, 152, 157, 161, 167, 182, 193, 196
healer, 2, 13, 17, 74, 75, 76, 126, 169, 170, 171, 172, 174, 175, 179, 181, 187, 188
healers, 14, 27, 74, 75, 80, 87, 155, 167, 169, 170, 188
healing, 2, 3, 6, 8, 12, 13, 14, 15, 16, 17, 22, 23, 24, 26, 30, 67, 68, 74, 76, 87, 91, 92, 95, 118, 119, 129, 132, 140, 141, 146, 150, 151, 153, 154, 155, 156, 157, 158, 160, 164, 167, 169, 170, 171, 173, 174, 175, 176, 178, 179, 183, 187, 188, 201
Healing, 1, 13, 25, 26, 27, 33, 76, 89, 91, 132, 143, 151, 153, 187, 197, 198, 199
healing process, 6, 23, 30, 158, 160, 169, 171, 179
health, 1, 2, 3, 5, 6, 7, 9, 12, 15, 19, 20, 21, 22, 27, 28, 29, 34, 35, 38, 39, 40, 43, 48, 49, 52, 58, 67, 70, 76, 94, 100, 108, 115, 123, 136, 137, 138, 139, 140, 141, 144, 146, 147, 149, 152, 155, 172, 174, 181, 182, 183, 190, 192, 193, 194, 195, 196, 200, 201
Hinduism, 79, 154
hippocampus, 116, 143, 180
Hippocrates, 8, 17, 146, 181
Hippocrates', 8
holistic, 7, 8, 16, 30, 33, 63, 72, 95, 96, 119, 127, 129, 138, 139, 143, 146, 173
homeostasis, 28, 60, 61, 65, 126, 137, 146
hope, 16, 29, 66, 181, 189
hormone replacement therapy, 32
Hormone therapies, 35
hormones, 32, 41, 54, 56, 65, 66, 71, 95, 110, 113, 115, 116, 119, 120, 127, 143, 144, 200
human condition, 27
Human Genome Project, 20
humanity, 17, 40, 41, 43, 62, 63, 125, 154, 170
humans, 40, 41, 42, 47, 50, 99, 101, 157, 171, 173, 187
humbleness, 12, 178
hypertension, 35
hypothalamus, 109, 116, 118, 124, 127
illness, 2, 7, 12, 14, 15, 19, 21, 22, 24, 25, 29, 48, 50, 88, 90, 91, 94, 96, 111, 121, 123, 126, 128, 135, 136, 137, 138, 139, 140, 141, 144, 146, 147, 148, 149, 150, 152, 173, 174, 175, 179, 183
imbalances, 23, 34, 116
immune system, 8, 14, 15, 25, 28, 33, 50, 51, 60, 70, 102, 106, 119, 120, 121, 122, 123, 124, 125, 127, 128, 129, 130, 132, 147, 150, 188
individualism, 43

industrialization, 36, 40, 48
infection, 14, 15, 25, 50, 60, 61, 129, 130
inflammation, 33, 89, 129, 130, 131, 160, 192, 193
information, 4, 6, 11, 12, 16, 17, 20, 26, 34, 51, 55, 59, 60, 65, 68, 70, 71, 78, 81, 86, 87, 93, 94, 97, 98, 112, 113, 114, 117, 118, 119, 123, 124, 125, 126, 135, 143, 153, 154, 161, 163, 164, 171, 172, 175, 176, 177, 180, 182, 189
initiations, 7, 161
inner healer, 1, 2, 8, 17, 18, 19, 21, 22, 30, 33, 37, 43, 60, 63, 66, 68, 69, 91, 95, 103, 109, 125, 126, 127, 133, 139, 141, 144, 147, 148, 161, 181
insurance, 29, 49
integrative medicine, 7, 141
integrity, 1, 18, 23, 26, 44, 174
intelligence, 13, 15, 18, 22, 26, 31, 61, 66, 75, 88, 103, 117, 126, 128, 159, 179
interpersonal relationships, 18, 26, 178
intuition, 11, 13, 14, 17, 31, 38, 63, 78, 85, 157, 166, 170, 171, 176
inventory, 26, 126
Kirlian photography, 74
knowledge, 1, 12, 13, 17, 19, 21, 26, 27, 29, 38, 49, 62, 79, 83, 89, 119, 141, 146, 154, 158, 161, 162, 163, 166, 168, 170, 175
Kybalion, 11, 13, 64, 164
Kyoto protocol, 45, 46
laws of life, 42, 43
laws of nature, 19
laying on of hands, 3, 4, 15, 17, 24, 87, 174
Learning, 26

lifestyle, 8, 9, 15, 18, 19, 22, 27, 37, 59, 62, 111, 141, 150, 156, 167, 174, 181, 189, 191, 194
limitation, 12, 22, 30, 92, 131, 134, 143
lineal, 63, 64, 72, 92, 113, 119
love, 8, 16, 41, 54, 78, 81, 83, 84, 85, 141, 150, 156, 157, 169, 171, 173, 195
loving our neighbors, 43
LSD, 115
lymph, 70, 102, 103, 109, 122, 132
magnetic, 73, 74, 75, 76, 77, 78, 98, 188
malnourishment, 39
master, 2, 4, 5, 6, 11, 117, 126, 152, 153, 156, 160, 161, 162, 163, 166, 168, 169, 175, 179
Master, 3, 1, 2, 3, 5, 58, 161, 187
mechanistic, 63
medical errors, 7, 94
medical treatment, 25, 136, 189
medication, 24, 31, 32, 35, 70, 114, 116, 126, 132, 134, 136, 140, 178, 189, 192
memory, 5, 62, 77, 113, 117, 143, 180
message, 34, 68, 69, 70, 91, 109, 110, 134, 135, 150
messages, 34, 68, 69, 101, 112, 113, 116, 126
messengers, 1, 32, 34, 68, 97, 109, 110, 112, 113, 119, 123, 126, 130
metabolism, 19, 50, 51, 55, 61, 69, 106, 109, 120, 130
molecules, 32, 50, 51, 67, 68, 71, 86, 87, 90, 92, 105, 107, 112, 123, 128, 130, 135, 183
morphine, 32
movement, 1, 4, 30, 35, 49, 74, 85, 92, 93, 96, 97, 100, 101, 108, 132, 141, 179, 180

206

multidimensional, 3, 1, 17, 19, 33, 52, 63, 68, 71, 72, 78, 79, 96, 114, 118, 140, 147, 150, 152, 159, 174, 176, 179, 183

multidimensionality, 22, 63, 71, 144

muscles, 30, 51, 56, 73, 92, 93, 98, 99, 100, 101, 103, 112, 127, 133, 134

nadis, 67, 70, 79, 86, 200

nervous system, 4, 33, 51, 56, 77, 78, 92, 97, 98, 99, 101, 103, 106, 108, 112, 113, 114, 115, 117, 121, 123, 124, 128, 180, 188

neurotransmitters, 112, 113, 115, 116, 119, 123

nutrigenomics, 19

nutrition, 15, 19, 21, 22, 24, 33, 34, 35, 38, 41, 56, 67, 70, 83, 100, 128, 133, 138, 144, 150, 174, 178, 179, 181, 190, 192, 195, 201

nutritional, 2, 15, 18, 19, 29, 31, 34, 36, 38, 39, 40, 41, 54, 130, 141, 189, 190, 193, 194, 195, 196

nutritional habits,, 18, 19, 141

obesity, 19, 35, 36, 39, 40, 54, 193, 194, 196

opiate, 68

optimal functioning, 49, 70, 109, 190

organic products, 20

Overweight, 35

ovulation, 65

pain, 2, 21, 22, 28, 30, 33, 50, 70, 75, 87, 88, 89, 91, 92, 99, 100, 105, 112, 116, 118, 129, 130, 131, 132, 133, 134, 135, 136, 140, 142, 143, 152, 154, 159, 160, 165, 173, 179, 188

pain killers, 2, 21, 70, 88, 134, 136

palliative, 25

pendulum, 4, 64, 85, 86

peptides, 68, 113, 123, 124

perception, 12, 57, 58, 63, 78, 83, 85, 113, 115, 117, 128, 138, 140, 141, 144, 152, 173, 178

perfection, 5, 27, 54, 173

personal transformation, 24

philosophy, 37, 63

physical activity, 22, 34, 66, 93, 100, 116, 127, 128, 144, 174, 179, 190, 192, 193

physical dimension, 25, 33, 81

pineal gland, 115, 120

pituitary, 65, 66, 109, 110, 115, 116, 118, 120, 123, 124, 125

Pleasure, 41

polarity, 64

practitioners, 11, 12, 13, 16, 17, 76, 78, 127, 135, 141, 151, 155, 158, 165, 166, 167, 170, 174, 189

prescription, 3, 4, 5, 7, 28, 29, 38, 88, 94

prevention, 22, 25, 29, 48, 54, 88, 190

Prevention, 22

privatization, 7, 9, 22, 48, 49

progress, 1, 27, 38, 40, 43, 44, 59, 167, 170, 174

psychology, 80, 177

psychotherapy, 25, 26, 127, 149

QiGong, 49

Quantum physics, 73

Quantum theory, 64

radiotherapy, 21

receptors, 32, 34, 52, 68, 70, 101, 108, 110, 113, 116, 134, 188, 200

refined products, 24, 41

reiki, 6, 7, 8, 11, 12, 13, 15, 16, 17, 22, 23, 26, 27, 30, 34, 49, 70, 76, 83, 85, 87, 89, 91, 126, 128, 132, 134, 135, 136, 151, 152, 153, 154, 155, 156, 157, 158,

159, 160, 161, 163, 164, 165, 166, 167, 169, 170, 174, 176, 179, 181, 187, 188, 189, 201

reiki masters, 11, 153, 167

reiki practitioner, 114, 188

relaxation, 6, 70, 75, 128, 132, 133, 134, 135, 152, 161, 167, 188

Religions, 63

resonance, 76, 78

rest, 28, 34, 36, 69, 89, 92, 117, 122, 133, 137, 139, 156, 180, 181

rhythmic life, 2, 166, 179, 180

sanitary infrastructure, 22, 48

science, 1, 3, 8, 21, 26, 28, 36, 41, 63, 64, 68, 72, 77, 90, 96, 98, 105, 111, 112, 118, 121, 122, 123, 125, 126, 130, 141, 144, 145, 146, 147, 168, 179, 193, 198, 199, 200

secretion, 33, 41, 104, 110, 115, 120, 124, 135

self esteem, 18

self-care, 49, 180

self-esteem, 27, 36, 144

serotonin, 66, 113, 114, 115, 135, 136

soul, 3, 27, 81, 82, 83, 117, 144, 158, 159, 176, 183

souls, 27, 170, 171, 176

spirit, 17, 25, 81, 86, 104, 117, 148, 164, 176, 177, 178

spiritual, 5, 6, 18, 19, 26, 67, 71, 79, 81, 82, 83, 84, 96, 114, 116, 117, 139, 144, 150, 151, 152, 153, 156, 157, 158, 162, 166, 167, 168, 170, 171, 174, 179

SQUID, 74, 76, 77

stress, 1, 18, 19, 22, 24, 33, 34, 43, 50, 52, 55, 56, 57, 59, 61, 71, 75, 100, 115, 116, 123, 126, 127, 128, 130, 138, 139, 142, 143, 144, 147, 148, 174, 179, 181, 188, 189, 192, 194

Stress, 18, 50, 144

stressors, 1, 18, 19, 52, 56, 57, 59, 96, 116, 128, 138, 139, 144

stretching, 92, 133

structure, 1, 8, 20, 69, 79, 82, 83, 86, 88, 89, 92, 93, 94, 98, 99, 100, 101, 102, 107, 116, 117, 123, 142, 161, 162

Structure, 92

students, 7, 11, 12, 16, 104, 119, 143, 149, 152, 153, 155, 157, 158, 160, 161, 162, 163, 164, 166, 187

subsonic, 75

subtle dimensions, 71, 96

suffering, 14, 18, 21, 28, 29, 34, 41, 61, 68, 79, 150, 174, 175

sugar, 20, 23, 24, 25, 38, 39, 44, 54, 101, 106, 107, 108, 112, 114, 146, 150, 181, 191, 193, 195

Suicide, 37, 198

symbols, 7, 11, 89, 119, 160, 161, 165, 166

symptom, 9, 13, 14, 16, 24, 25, 26, 116, 134, 135, 137, 138, 140, 143, 159, 174, 183

symptoms, 8, 13, 14, 15, 19, 24, 25, 29, 32, 37, 49, 55, 61, 70, 87, 89, 90, 100, 118, 119, 126, 127, 136, 138, 139, 140, 142, 144, 147, 148, 150, 151, 166, 167, 178, 195, 198

systemic epistemology, 64

Tao, 64

Taoism, 79, 80

technology, 19, 21, 38, 41, 42, 47, 66, 94, 172

thalamus, 123, 135

the Earth, 41, 42, 77, 145, 146, 183, 188

therapeutic, 2, 14, 77, 105, 146, 171, 177, 178, 181, 189

therapist, 16, 70, 76, 93, 171, 172, 177, 178, 181

Therapy, 16, 87, 198, 199

third world 22, 35, 40, 41, 48

thyroid, 1, 2, 66, 95, 120, 144

toxicity, 19, 61, 137

Traditional Chinese Medicine, 2, 67, 79, 151

trauma, 32, 57, 134, 142, 143, 159, 160

traumatic, 18, 19, 126, 142, 143

treatments, 8, 13, 14, 15, 25, 35, 85, 92, 108, 141, 142, 160, 166, 167, 175, 179, 189, 192

unified opposites, 92

United States, 22, 34, 35, 37, 44, 45, 46, 47, 49, 52, 53, 54, 75, 94, 138, 141, 146, 153

universal energy, 16, 67, 80, 83, 85, 156, 167

Universe, 1, 8, 17, 37, 52, 64, 183

urbanization, 36, 48

vegetables, 33, 36, 106, 109, 131, 190, 191, 193

vegetarian, 3, 5, 6, 189

vibration, 72, 75, 76, 78, 83, 143, 164

vibrational, 7, 63, 75, 80, 86, 87, 139, 165

vitamins and minerals, 28, 34

wellbeing, 28, 67, 139, 140

wellness, 24, 94, 141

wholeness, 24, 27, 41, 63, 91, 167, 174, 188

willpower, 26, 152

Ying and Yang, 64

www.ingramcontent.com/pod-product-compliance
Lightning Source LLC
Chambersburg PA
CBHW031928190326
41519CB00007B/452